Medical Equipment Maintenance: Management and Oversight

Synthesis Lectures on Biomedical Engineering

Editor
John D. Enderle, ***University of Connecticut***

Lectures in Biomedical Engineering will be comprised of 75- to 150-page publications on advanced and state-of-the-art topics that spans the field of biomedical engineering, from the atom and molecule to large diagnostic equipment. Each lecture covers, for that topic, the fundamental principles in a unified manner, develops underlying concepts needed for sequential material, and progresses to more advanced topics. Computer software and multimedia, when appropriate and available, is included for simulation, computation, visualization and design. The authors selected to write the lectures are leading experts on the subject who have extensive background in theory, application and design.
The series is designed to meet the demands of the 21st century technology and the rapid advancements in the all-encompassing field of biomedical engineering that includes biochemical, biomaterials, biomechanics, bioinstrumentation, physiological modeling, biosignal processing, bioinformatics, biocomplexity, medical and molecular imaging, rehabilitation engineering, biomimetic nano-electrokinetics, biosensors, biotechnology, clinical engineering, biomedical devices, drug discovery and delivery systems, tissue engineering, proteomics, functional genomics, molecular and cellular engineering.

Medical Equipment Maintenance: Management and Oversight
Binseng Wang
2012

Capstone Design Courses, Part II: Preparing Biomedical Engineers for the Real World
Jay R. Goldberg
2012

Chronobioengineering: Introduction to Biological Rhythms with Applications
Donald McEachron
2012

Ethics for Bioengineers
Monique Frize
2011

Computational Genomic Signatures
Ozkan Ufuk Nalbantoglu and Khalid Sayood
2011

Digital Image Processing for Ophthalmology: Detection of the Optic Nerve Head
Xiaolu Zhu, Rangaraj M. Rangayyan, and Anna L. Ells
2011

Modeling and Analysis of Shape with Applications in Computer-Aided Diagnosis of Breast Cancer
Denise Guliato and Rangaraj M. Rangayyan
2011

Analysis of Oriented Texture with Applications to the Detection of Architectural Distortion in Mammograms
Fábio J. Ayres, Rangaraj M. Rangayyan, and J. E. Leo Desautels
2010

Fundamentals of Biomedical Transport Processes
Gerald E. Miller
2010

Models of Horizontal Eye Movements, Part II: A 3rd Order Linear Saccade Model
John D. Enderle and Wei Zhou
2010

Models of Horizontal Eye Movements, Part I: Early Models of Saccades and Smooth Pursuit
John D. Enderle
2010

The Graph Theoretical Approach in Brain Functional Networks: Theory and Applications
Fabrizio De Vico Fallani and Fabio Babiloni
2010

Biomedical Technology Assessment: The 3Q Method
Phillip Weinfurt
2010

Strategic Health Technology Incorporation
Binseng Wang
2009

Phonocardiography Signal Processing
Abbas K. Abbas and Rasha Bassam
2009

Introduction to Biomedical Engineering: Biomechanics and Bioelectricity - Part II
Douglas A. Christensen
2009

Introduction to Biomedical Engineering: Biomechanics and Bioelectricity - Part I
Douglas A. Christensen
2009

Landmarking and Segmentation of 3D CT Images
Shantanu Banik, Rangaraj M. Rangayyan, and Graham S. Boag
2009

Basic Feedback Controls in Biomedicine
Charles S. Lessard
2009

Understanding Atrial Fibrillation: The Signal Processing Contribution, Part I
Luca Mainardi, Leif Sörnmo, and Sergio Cerutti
2008

Understanding Atrial Fibrillation: The Signal Processing Contribution, Part II
Luca Mainardi, Leif Sörnmo, and Sergio Cerutti
2008

Introductory Medical Imaging
A. A. Bharath
2008

Lung Sounds: An Advanced Signal Processing Perspective
Leontios J. Hadjileontiadis
2008

An Outline of Informational Genetics
Gérard Battail
2008

Neural Interfacing: Forging the Human-Machine Connection
Susanne D. Coates
2008

Quantitative Neurophysiology
Joseph V. Tranquillo
2008

Tremor: From Pathogenesis to Treatment
Giuliana Grimaldi and Mario Manto
2008

Introduction to Continuum Biomechanics
Kyriacos A. Athanasiou and Roman M. Natoli
2008

The Effects of Hypergravity and Microgravity on Biomedical Experiments
Thais Russomano, Gustavo Dalmarco, and Felipe Prehn Falcão
2008

A Biosystems Approach to Industrial Patient Monitoring and Diagnostic Devices
Gail Baura
2008

Multimodal Imaging in Neurology: Special Focus on MRI Applications and MEG
Hans-Peter Müller and Jan Kassubek
2007

Estimation of Cortical Connectivity in Humans: Advanced Signal Processing Techniques
Laura Astolfi and Fabio Babiloni
2007

Brain–Machine Interface Engineering
Justin C. Sanchez and José C. Principe
2007

Introduction to Statistics for Biomedical Engineers
Kristina M. Ropella
2007

Capstone Design Courses: Producing Industry-Ready Biomedical Engineers
Jay R. Goldberg
2007

BioNanotechnology
Elisabeth S. Papazoglou and Aravind Parthasarathy
2007

Bioinstrumentation
John D. Enderle
2006

Fundamentals of Respiratory Sounds and Analysis
Zahra Moussavi
2006

Advanced Probability Theory for Biomedical Engineers
John D. Enderle, David C. Farden, and Daniel J. Krause
2006

Intermediate Probability Theory for Biomedical Engineers
John D. Enderle, David C. Farden, and Daniel J. Krause
2006

Basic Probability Theory for Biomedical Engineers
John D. Enderle, David C. Farden, and Daniel J. Krause
2006

Sensory Organ Replacement and Repair
Gerald E. Miller
2006

Artificial Organs
Gerald E. Miller
2006

Signal Processing of Random Physiological Signals
Charles S. Lessard
2006

Image and Signal Processing for Networked E-Health Applications
Ilias G. Maglogiannis, Kostas Karpouzis, and Manolis Wallace
2006

© Springer Nature Switzerland AG 2022
Reprint of original edition © Morgan & Claypool 2012

All rights reserved. No part of this publication may be reproduced, stored in a retrieval system, or transmitted in any form or by any means—electronic, mechanical, photocopy, recording, or any other except for brief quotations in printed reviews, without the prior permission of the publisher.

Medical Equipment Maintenance: Management and Oversight
Binseng Wang

ISBN-13: 978-3-031-00527-5 paperback
ISBN-13: 978-3-031-01655-4 ebook

DOI 10.1007/978-3-031-01655-4

A Publication in the Springer series
SYNTHESIS LECTURES ON BIOMEDICAL ENGINEERING

Lecture #45
Series Editor: John D. Enderle, *University of Connecticut*
Series ISSN
Synthesis Lectures on Biomedical Engineering
Print 1930-0328 Electronic 1930-0336

Medical Equipment Maintenance: Management and Oversight

Binseng Wang
ARAMARK Healthcare Technologies

SYNTHESIS LECTURES ON BIOMEDICAL ENGINEERING #45

ABSTRACT

In addition to being essential for safe and effective patient care, medical equipment also has significant impact on the income and, thus, vitality of healthcare organizations. For this reason, its maintenance and management requires careful supervision by healthcare administrators, many of whom may not have the technical background to understand all of the relevant factors. This book presents the basic elements of medical equipment maintenance and management required of healthcare leaders responsible for managing or overseeing this function. It will enable these individuals to understand their professional responsibilities, as well as what they should expect from their supervised staff and how to measure and benchmark staff performance against equivalent performance levels at similar organizations.

The book opens with a foundational summary of the laws, regulations, codes, and standards that are applicable to the maintenance and management of medical equipment in healthcare organizations. Next, the core functions of the team responsible for maintenance and management are described in sufficient detail for managers and overseers. Then the methods and measures for determining the effectiveness and efficiency of equipment maintenance and management are presented to allow performance management and benchmarking comparisons. The challenges and opportunities of managing healthcare organizations of different sizes, acuity levels, and geographical locations are discussed. Extensive bibliographic sources and material for further study are provided to assist students and healthcare leaders interested in acquiring more detailed knowledge.

KEYWORDS

medical equipment, clinical engineering, biomedical engineering, healthcare technology management, equipment maintenance and management, benchmarking, maintenance effectiveness, evidence-based maintenance

Contents

Preface

This book is an attempt to provide the basic elements of medical equipment maintenance for those who have the responsibility to manage or oversee this function in healthcare organizations and systems, especially those who may not have the technical background or experience in this specialized area, such as senior healthcare executives (hospital COO, CFO, etc., VP of Support Services, Supply Chain, or Information Technology), as well as hospital administrators and students of healthcare administration, biomedical engineering, or business management. It can also be helpful to new clinical engineering managers and biomedical engineering technicians who aspire to career growth in management.

The material presented here is based on over 30 years of experience in this field, both in the U.S. and in other countries where I worked or provided consulting services. Obviously, I benefited from the extensive work of those who pioneered the field, as well as from my contemporaries who have been trying to take this profession to a higher level. In particular, I would like to acknowledge the contributions of my current colleagues and friends at ARAMARK Healthcare Technologies and former colleagues and friends at MEDIQ/PRN Life Support Services, National Institutes of Health, Secretariat of Health of Sao Paulo state and State University of Campinas in Brazil, Pan-American Health Organization, and World Health Organization, whose knowledge, experience, and feedback have helped me to improve continually my understanding of the challenges of managing medical equipment maintenance. I apologize in advance if I failed to cite a specific work or acknowledge someone but request feedback to allow me to make corrections. On the other hand, I must assume full responsibilities for any shortcomings in this book. If I learned anything over these years is that the most important and humbling lessons were learned from my own mistakes. Many of the concepts and processes outlined here may seem simple and obvious to the reader but often took a long time for me to realize that these are the right things to do, in the right way, and at the right time.

For this reason, I do not pretend that the material presented in this book is complete, correct, or the best of what is known in this field but only what I believe is a rational and organized approach to the management of medical equipment maintenance. I sincerely hope that the readers can take advantage of what is presented here to develop their own and better practice and, thus, contribute to better use of resources to enhance reliability and safety of medical equipment for the benefit of patients worldwide. Criticisms, comments, suggestions, and new ideas are most welcome.

Binseng Wang
September 2012

CHAPTER 1

Introduction

Medical equipment is extensively used in all aspects of health services, ranging from prevention, screening, diagnosis, monitoring, and therapeutics to rehabilitation. Nowadays, it is virtually impossible to provide health services without it. Unlike other types of healthcare technologies (i.e., drugs, implants, and disposable products), medical equipment requires maintenance (both scheduled and unscheduled) during its useful life. As the sophistication and cost of medical equipment continue to escalate, the complexity and cost of its maintenance have also risen sharply in the last few decades.

Studies conducted using data collected from hundreds of acute-care hospitals indicate that on average, each hospital acquired about 15–20 pieces of medical equipment for each staffed bed, which translates into a capital investment of around US$200–400,000/staffed bed [Wang et al., 2008, 2012]. Thus, it is common for a 500-bed hospital to own more than US$100–200 million worth of medical equipment and considerably more if it is affiliated with a medical school. The same studies have indicated that annual medical equipment maintenance and management cost is approximately 1% of the total hospital budget, so a 500-bed hospital spends typically around $5 million/year.

In addition to its high maintenance costs, medical equipment is often involved in patient incidents that resulted in serious injuries or deaths. In fact, statistics accumulated by the Joint Commission (TJC) show medical equipment-related "sentinel events[1]" is typically among the top ten types every year [TJC, 2012a]. Although root-cause analyses of these incidents show that the majority of them are not caused by equipment failures or maintenance omissions [Wang et al., in press], it is still the hospital's leadership responsibility to address them, as the primary causes are human factors, leadership, communication, and clinical assessment [TJC, 2012a]. Experts agree that these causes are best addressed through user training and enhanced interaction between clinicians and technical staff until more comprehensive improvements can be achieved with the involvement of regulatory agencies and manufacturers [AAMI, 2010a, 2011a].

For these reasons, every hospital now needs a team of highly qualified technical personnel to maintain and manage the equipment during its lifetime. This technical team was initially and still called in many hospitals the "biomed" department as a short hand for biomedical engineering. Since the early 1990s, many hospitals have renamed it to clinical engineering (CE), as biomedical engineering is nowadays typically understood as a research and teaching department that focuses on the design and manufacturing of medical devices and the study of biological systems. The American

[1] Sentinel event is defined by TJC as "an unexpected occurrence involving death or serious physical or psychological injury, or the risk thereof. Serious injury specifically includes loss of limb or function. The phrase 'or risk thereof' includes any process variation for which a recurrence would carry a significant chance of a serious adverse outcome."

College of Clinical Engineering (ACCE) defines clinical engineer as "a professional who supports and advances patient care by applying engineering and managerial skills to healthcare technology" [ACCE, 2012]. Another term that was used was "asset management" but it did not catch on. More recently, some have recommended the adoption of healthcare technology management (HTM), although this could create some confusion with those who work with information technology (IT) in healthcare or material management [AAMI, 2011b]. In the remainder of this book, the term CE will be used.

Like other non-clinical departments within the hospital, the CE Department may report to Facility Management, Material Management, Support Services, Administration, or Information Technology. Regardless of the name of the overseeing department and the title of its leader, the management and oversight of CE is not a trivial task. While the availability of a highly qualified CE manager can alleviate the responsibilities of whoever is responsible for the oversight, it is still highly desirable that the overseer not only know what the CE Department is doing but also why, as well as how to measure its performance and guide it to continually improve its operations. This book is written for managers and overseers of CE departments who have limited prior CE experience and would like to gain good command of the regulatory requirements, core functions, performance measures, and interactions with other clinical and non-clinical departments. This book can also serve as a brief overview of medical equipment maintenance and management for those who are intrigued by the profession and would like to gain some knowledge before committing to a career in this field. Finally, this book could also be a valuable tool for biomedical engineering technicians (BMETs) who would like to become CE managers in detecting areas of study that they need to focus on before becoming ready for career growth.

Before describing the core functions of the CE Department, a review of federal, state and local laws, regulations, and codes is made to provide the regulatory framework of CE activities. Next, each of the main CE responsibilities are described, starting with equipment planning and acquisition, progressing through acceptance, technical training of clinical staff, scheduled and unscheduled maintenance, management of vendor services, assistance in safety and risk management, supervision of technical staff, internal financial management, and ending with replacement, retirement, and disposal of undesired equipment. Internal management of the CE Department, i.e., staff management and education, process and record management, financial management, performance monitoring, and continual improvement, will also be covered. As this book is not meant to be a cookbook recipe but a set of generic guidelines, discussion of implementation alternatives and challenges is provided to help managers and overseers make difficult decisions. Although almost all the information provided was collected within the U.S., foreign readers can still benefit from the material by making suitable adjustments for their respective environments.

CHAPTER 2

Regulatory Framework

Healthcare is highly regulated in the U.S. so it should not be surprising that there are a myriad of laws, regulations, and codes that are applicable to the maintenance and management of medical equipment in almost any care delivery site. A summary is provided below without the intent of being exhaustive due to space limitation. Also, emphasis is given to acute care hospitals because they form the majority of hospitals and own most equipment. However, most of the regulatory requirements mentioned here also apply, with some differences, to clinics, ambulatory surgical centers, long-term care sites, etc. Besides federal and state requirements, some applicable standards are also summarized.

2.1 FEDERAL LAWS, REGULATIONS, AND CODES

Once a law is created by Congress and signed by the president, a government agency assigned by the law is responsible for its implementation. This is accomplished through the issuance of a regulation that has specific requirements and penalties. Table 2.1 shows the main federal laws and regulations, with respective enforcing agency, that are applicable to medical equipment maintenance and management. Each of these requirements is outlined briefly below under the respective enforcing agency, i.e., the Centers for Medicare & Medicaid Services (CMS), the Food and Drug Administration (FDA), and the Office for Civil Rights (OCR), which are part of the Department of Health and Human Services (HHS), the Occupational Safety and Health Administration (OSHA) which is under the Department of Labor (DOL), and the Federal Emergency Management Agency (FEMA) which belongs to the Department of Homeland Security (DHS).

2.1.1 REGULATIONS ENFORCED BY CMS

Conditions of Participation

The Conditions of Participation (CoPs) are only required for healthcare organizations that are registered with CMS as "providers," and, thus, its compliance is voluntary in theory. In practice, it is mandatory for the vast majority of American hospitals, as approximately 47% of hospital care expenditures in the U.S. are funded by Medicare and Medicaid reimbursements [CMS, 2012a]. Very few hospitals can afford to forgo those reimbursements; furthermore, private insurance companies are unlikely to be willing to enlist hospitals not registered with CMS to provide services to its insurance policy holders.

The CoP applicable to medical equipment maintenance and management is limited to one single clause 42CFR482.41(c)(2) which states: "Facilities, supplies, and equipment must be maintained to ensure an acceptable level of safety and quality." This CoP has been interpreted by CMS

Table 2.1: Federal laws and regulations applicable to medical equipment maintenance and management with the respective agencies empowered to enforce them (see text for their full names). The role of accreditation organizations and state agencies is explained in the text.

LAW	REGULATION	ENFORCING AGENCY
Social Security Amendments 1965 Medicare Improvements for Patients and Providers Act 2008	42 CFR 482: Conditions of Participation	CMS (HHS), Accreditation Organizations, and State Agencies
Clinical Lab Improvement Amendments (CLIA) 1988	42 CFR 493: Laboratory Requirements	CMS (HHS), Accreditation Organizations
Health Insurance Portability & Accountability Act (HIPAA) 1996 Patient Protections and Affordable Care Act 2010	45 CFR 160, 162 & 164: Privacy Rule 45 CFR 160, 162 & 164: Security Rule	Office for Civil Rights (HHS)
Food & Drug Act 1906 Medical Device Amendment 1976	21 CFR 606: Human blood and blood products 21 CFR 801: Labeling 21 CFR 803: Medical device reporting 21 CFR 806: Medical devices; reports of corrections and removals 21 CFR 810: Medical device recall authority 21 CFR 820: Quality system regulation 21 CFR 821: Medical device tracking requirements 21 CFR 880.6310: Medical device data system 21 CFR 1000 – 1050: Radiation Health	FDA (HHS)
Occupational Safety and Health Act 1970	29 CFR 1910: Occupational Safety and Health Standards	OSHA (DOL)
Disaster Mitigation Act of 2000 Homeland Security Act of 2002 Disaster Relief and Emergency Assistance Act of 1988	44 CFR subchapter F: Preparedness	FEMA (DHS)

until 2011 as "equipment shall be maintained according to manufacturers' recommendations" [CMS, 2011]. This interpretation differs sharply from the Environment of Care (EOC) standards issued and enforced by the Joint Commission (TJC) [TJC, 2012b]. In the last almost 25 years, TJC gradually transitioned from rigid maintenance requirements to a more flexible approach in which hospitals are allowed to determine their own maintenance strategies based on a combination of risk assessment, maintenance experience, and manufacturers' recommendations [Ridgway, 2004; Stiefel, 2009].

The discrepancy between CMS interpretations and TJC practice came to CMS attention in 2009 soon after the Medicare Improvements for Patients and Providers Act 2008 gave CMS the power to oversee all accreditation organizations[2]. In December 2011, CMS issued a clarification letter with a revision to its interpretation of the CoP allowing flexibility for hospitals to adopt less frequent maintenance schedules for non-critical equipment but still requires them to follow manufacturers' maintenance procedures. At the time of the publication of this book, the revised interpretation is still considered excessively stringent by the professionals involved in equipment maintenance and management, so this issue remains unresolved.

Since CMS does not have resources to enforce CoP by itself, it relies on "accreditation organizations" (AOs) to survey hospitals that participate in the Medicare and Medicaid programs. In addition, CMS requires states to use its own agencies to conduct audits of hospitals for validation of surveys conducted by AOs, as well as to investigate complaints filed by patients and their families. Three AOs have been authorized by CMS for hospital accreditation:

- American Osteopathic Association (AOA)
- DNV Healthcare, a division of Det Norske Veritas
- The Joint Commission (TJC)

In addition, CMS has four other approved AOs for survey of ambulatory surgical centers, hospitals, home care, rehabilitation facilities, and rural health clinics.

Clinical Laboratory Requirements

All clinical laboratories, freestanding or affiliated with the hospital, that perform even one test, including waived tests, on "materials derived from the human body for the purpose of providing information for the diagnosis, prevention, or treatment of any disease or impairment of, or the assessment of the health of, human beings" are required to fulfill certain quality standards per the Clinical Laboratory Improvement Amendments of 1988 (CLIA). The CLIA program is managed financially by CMS, which authorizes certain organizations to perform the inspections, and supported technically by the FDA.

Within the CLIA program, there are specific requirements on the maintenance and functional checks of clinical laboratory equipment (42 CFR 493.1215), as well as for calibration and

[2] Until 2008, TJC and the American Osteopathic Association (AOA) were granted "deemed status" to accredit hospitals by the Social Security Amendments of 1965.

calibration verification (42 CFR 493.1217). The CE Department will assist the clinical laboratory manager in complying with CLIA and other applicable standards and regulations by maintaining equipment using protocol and frequency recommended by the respective manufacturer. If protocols or frequencies are not provided by the manufacturer, the equipment, instruments, or test systems were developed in-house or commercially available but modified by the laboratory, CE staff will follow the maintenance protocol and frequency established by the laboratory itself. If the equipment is leased from vendors (such as part of a reagent purchase agreement), the respective vendor will be required to provide documents to prove that it has been maintaining equipment according to CLIA requirements.

CMS has approved six accreditation organizations for survey of laboratories that perform "nonwaived" (moderate and/or high complexity) testing[3]:

- American Association of Blood Banks (AABB)
- American Osteopathic Association (AOA)
- American Society of Histocompatibility and Immunogenetics (ASHI)
- COLA
- College of American Pathologists (CAP)
- The Joint Commission (TJC)

2.1.2 REGULATIONS ENFORCED BY OCR

HIPAA Security Rule

Hospitals (as well as other healthcare organizations, insurance companies, and claims processing agencies) are required to adopt processes and tools to safeguard *protected health information* (PHI) by the Health Insurance Portability and Accountability Act of 1996 (HIPAA). These processes and tools include administrative, physical, and technical safeguards to ensure the confidentiality, integrity, and security of electronic PHI.

While the primary responsibility of HIPAA Security Rule compliance responsibility falls on the Information Technology (IT) department, the CE Department has to use its expertise to assist IT by: (i) identifying which pieces of medical equipment contain PHI, (ii) evaluating the risk of unauthorized access to PHI, and (iii) suggesting means to protect PHI.

Currently, only a small fraction of medical equipment collects and stores patient information, so it is not necessary to perform a comprehensive study of every piece of medical equipment within the hospital. The trend is, however, more and more equipment will contain PHI due to the desire to record all patient related data within the electronic health records (EHR), sometimes also known as electronic medical records (EMR). Once the equipment that collects and stores patient

[3]CLIA defines "waived tests" as "simple laboratory examinations and procedures that have an insignificant risk of an erroneous result."

information has been identified, one can request from the respective manufacturer how the patient information is stored, transmitted, and protected. This information is commonly provided using the Manufacturer Disclosure Statement for Medical Device Security (MDS2) form that was developed in collaboration with a number of professional associations [HIMSS, 2012]. Using the MDS2 information, appropriate safeguards can be developed to protect PHI within medical equipment using a combination of passwords, isolation from networks, network protection, virus protection, firewalls, etc.

HIPPA Privacy Rule

The Privacy Rule complements the Security Rule by requiring appropriate safeguards to protect the privacy of PHI, and sets limits and conditions on the uses and disclosures that may be made of such information without patient authorization.

The CE Department is responsible for controlling PHI stored in medical equipment to prevent unauthorized use while the equipment is being serviced[4]. This includes controlling the access of medical equipment to service vendors, deleting or encrypting PHI stored in equipment sent out for service, and removing all PHI before disposing retired equipment. In addition, the CE Department needs to be trained to comply with the hospital's HIPAA policies and procedures.

In case CE is outsourced, the hospital may opt to require the outsourcing company to sign a *business associate* (BA) contract, thus transferring the compliance of Privacy Rules to the latter. According to the Office for Civil Rights' interpretation of the HIPAA Privacy Rule (45 CFR 164.502(a)(1)), the BA agreement is not mandatory, as any disclosure of PHI that occurs in the performance of their duties (such as what may occur while repairing a piece of medical equipment) is limited in nature, occurs as a by-product of the maintenance duties, and cannot be reasonably prevented [OCR, 2012]. Regardless of the existence of the BA agreement, it is strongly recommended that all third-party employees working in the hospital be required to receive the same training as hospital employees, even when the disclosure of PHI is incidental.

2.1.3 REGULATIONS ENFORCED BY FDA

While most of the medical device regulations are applicable to device manufacturers, some of them apply also to hospitals and, therefore, to CE departments.

Human Blood and Blood Products

While blood and blood products are not obviously medical equipment, the FDA regulations for blood banks and transfusion services have several specific requirements applicable to equipment used in these services. Therefore, if a hospital offers these services on site, it is subject to FDA inspections, and the CE Department must comply with equipment maintenance and management requirements.

[4]The protection of PHI during clinical use is the responsibility of clinical departments with assistance from IT.

The specific requirements for equipment used in the blood bank and transfusion service are spelled out in 21 CFR 606.60, which states "Equipment used in the collection, processing, compatibility testing, storage and distribution of blood and blood components shall be maintained in a clean and orderly manner and located so as to facilitate cleaning and maintenance. The equipment shall be observed, standardized and calibrated on a regularly scheduled basis as prescribed in the Standard Operating Procedures Manual and shall perform in the manner for which it was designed so as to assure compliance with the official requirements prescribed in this chapter for blood and blood products." In addition, it requires performance checks by operators and calibration of certain types of equipment.

However, in its Compliance Program Guidance Manual [FDA, 2011], FDA directs its inspectors to review some samples of maintenance procedures and records to verify that the establishment is following its Standard Operating Procedures and if those "procedures conform to the equipment manufacturer's recommendations." So in essence, manufacturers' recommendations have become requirements again.

Adulterated and Misbranded Medical Devices

FDA enforces the Food, Drug, and Cosmetic Act (FD&C) (21 USC 301) originally issued in 1906 and amended numerous times, including the Medical Device Amendments of 1976, which started the regulation of medical devices to ensure their safety and effectiveness. Two of the basic requirements for all medical devices (known as *General Controls*) are adulteration (section 501 of FD&C and regulated through 21 CFR 820) and misbranding (section 502 of FD&C and regulated through 21 CFR 801). Among other implications, these requirements prevent anyone other than a registered manufacturer or remanufacturer of medical devices to modify any equipment's technical specifications or clinical indications, as this action would constitute adulteration and misbranding, respectively.

It is, however, necessary to interpret carefully these regulations. For example, they do not prevent CE departments from performing scheduled maintenance and repairs, as long as the equipment specifications are not altered, including using parts and supplies that are not purchased directly from the manufacturer. Furthermore, CE staff can perform upgrades and updates approved and issued by the respective manufacturers. Cosmetic (e.g., painting) and functional (e.g., replacing existing casters with bigger ones to ease movements) changes that do not affect equipment specifications are also acceptable. On the other hand, any changes to the specifications or modifications to allow using the equipment on new clinical conditions cannot be performed without an Investigation Device Exemption (IDE) from the FDA, including approval of the hospital's Institutional Review Board.

Medical Device Recall

Recalls are issued by manufacturers or the FDA to address problems in equipment that can pose risks to health or violated FDA regulations (21 CFR 7, 80, 806 and 810). Recalls may be conducted on a manufacturer's own initiative, by FDA request, or by FDA order under statutory authority. The manufacturer can remove the product from the market or correct the violation. In the case of medical equipment, manufacturers typically make corrections—often called updates or upgrades—instead of removals because of the high cost of replacing the entire equipment. When a manufacturer identifies a problem and starts a correction process, it typically issues a "safety alert" and notifies the FDA. It is up to the FDA to designate the correction as a recall and classify it into one of the three categories listed below. This explains why there is often a significant delay between the initial receipt of an alert by the equipment owner and the subsequent publication and classification of the recall.

- **Class I recall:** a situation in which there is a reasonable probability that the use of or exposure to a violative product will cause serious adverse health consequences or death.

- **Class II recall:** a situation in which use of or exposure to a violative product may cause temporary or medically reversible adverse health consequences or where the probability of serious adverse health consequences is remote.

- **Class III recall:** a situation in which use of or exposure to a violative product is not likely to cause adverse health consequences.

While a recall does not mean that a piece of equipment can no longer be used, it is also not prudent to ignore a recall for two reasons. First, the FDA has the authority to seize recalled products and take legal action against those who refuse to comply. The second, and perhaps more important reason, is the hospital may be liable for a malpractice lawsuit if a patient is injured by a piece of equipment that was recalled and the recall was ignored.

For this reason, the CE Department has to be vigilant in monitoring recalls and managing them in a timely manner. Often the recall information is sent to the clinical department or material management only, so the CE Department needs to rely on alternative sources such as the FDA's website (`http://www.fda.gov/MedicalDevices/Safety/RecallsCorrectionsRemovals/ListofRecalls/default.htm`) and paid services such as *Biomedical Safety and Standards* published by Lippincott Williams & Wilkins, ECRI Institute's *Alerts Tracker*, and Noblis's *RASMAS* (see Appendix B—Information and Data Sources on the Internet). Tracking of recall completion is critical and should be included in the computerized maintenance management system (CMMS) used by CE for managing equipment (see Section 4.2 below).

Incident Investigation and Reporting

Hospitals are required by the Safe Medical Devices Act of 1990 (SMDA) to report suspected medical device related serious injuries to the manufacturer and deaths to both the manufacturer and FDA, as well as file an annual report to summarize their adverse event reports (21 CFR 803). In addition, some state departments of health or hospital licensing boards may also have similar requirements.

Before reporting any incident, the hospital must first decide whether the medical equipment was indeed involved and contributed to the injury or death, since numerous other factors may be involved. For this reason, it is critical to involve the CE Department to perform (or witness in case an independent third-party investigator is retained) a systematic and objective investigation of all incidents involving equipment, under the coordination of the hospital's Risk Manager. Appropriate methodology for conducting incident investigations, including precautions that can protect the hospital in potential litigations, has been described elsewhere and is too lengthy to be discussed here (see, e.g., [Geddes, 2002; Dyro, 2004]). Since the same methods can be used for medical devices not managed by the CE Department, CE staff can be helpful in those cases as well.

Medical Device Tracking

Certain medical devices (mostly implants and some moveable equipment used *outside* of hospitals) are required by the FDA to be tracked as to its current location and user per the medical device tracking requirement established by the Safe Medical Device Act of 1990 (SMDA) and modified by the FDA Modernization Act of 1997 and the FDA Amendments Act of 2007 (21 CFR 821).

If the hospital has a homecare division that provides the following kinds of equipment to patients to be used outside of the hospital, a tracking system must be created and managed:

- Breathing frequency monitors (e.g., apnea monitors)
- Continuous ventilators
- DC-defibrillators and paddles
- Ventricular bypass (assist) device

This tracking system does not need to be included in the CMMS used by the CE for managing equipment (see Section 4.2 below), but the CE Department must be informed of the constant changes in the location of these devices in order to plan for scheduled maintenance, monitor recalls, and update the inventory system. In particular, the acquisition of new devices and the disposal (including loss) of existing devices are critical pieces of information needed to keep the inventory data current and accurate.

Medical Device Data System

Medical Device Data System (MDDS) is defined by the FDA (21 CFR 880.6310) as "a device that is intended to provide one or more of the following uses, without controlling or altering the functions or parameters of any connected medical devices: (i) The electronic transfer of medical device data; (ii) The electronic storage of medical device data; (iii) The electronic conversion of medical device data from one format to another format in accordance with a preset specification; or (iv) The electronic display of medical device data." The FDA clarifies that MDDS "does not include devices intended to be used in connection with active patient monitoring."

Early in 2011 the FDA reclassified MDDS from class III (high risk) devices to class I (low risk), thus exempting the manufacturers from premarket review and approval[5]. However, MDDS

[5] See announcement at http://www.fda.gov/MedicalDevices/ProductsandMedicalProcedures/GeneralHospitalDevicesandSupplies/MedicalDeviceDataSystems/ucm251897.htm.

manufacturers, including hospitals that have developed their own MDDS, are required to register with the FDA, list their MDDS products, reporting adverse events, and comply with the FDA's Quality Systems regulation (21 CFR 820).

CE departments are, therefore, required to treat MDDS like other regulated devices in not adulterating or mislabeling them, performing recalls as required, and assisting in the investigation and reporting of patient incident. Most importantly, CE departments should refrain from developing hardware, software, and communication interfaces and protocols that are considered MDDS without proper a priori consultation with the hospital's regulatory authorities.

Radiation Protection

Hospitals are required to protect their employees and patients, as well as the environment, from excessive exposure to ionizing radiation. Several government agencies enforce specific regulations, depending on the protection target. The Environmental Protection Agency (EPA) is responsible for radiation contamination of the environment (air, water, etc.). The Occupational Safety and Health Administration (OSHA) is responsible for the protection of workers (and patients to some extent) through 29 CFR 1910.1096. The FDA is responsible for regulating manufacturers and installers of radiation emitting devices. Finally, the Nuclear Regulatory Commission sets the global standards for protection (10 CFR 20).

Typically, each hospital has a designated radiation safety officer. CE staff will work closely with the officer to protect patients, workers (including themselves), and the environment by keeping radiation exposure *"as low as reasonably achievable"* (ALARA), using the cardinal principles for radiation safety protection defined by the National Council on Radiation Protection (i.e., time, distance, and shielding). In addition, CE staff that maintains radiation-emitting equipment must report any exposure risks detected in the equipment to the radiation safety officer.

X-ray System Assembly Reporting

FDA regulation (21 CFR 1010.2 and 1020.30) requires the reporting of installation and reinstallation of X-ray systems and "non-certified" components[6] in such systems using a specific form (FDA 2579). Installers are required to follow the instructions provided by the respective manufacturer for assembly, installation, adjustment, and testing (AIAT) of X-ray systems. Such AIAT instructions are required by 21 CFR 1020.30(g) to be provided to the assembler upon request (at a cost not to exceed the cost of publication and distribution). Many state radiation protection agencies have similar requirements.

CE departments that install, reinstall (e.g., after moving the equipment), or repair X-ray producing systems need to read carefully the regulations to understand the requirements, as there are numerous exceptions. For example, reloaded or replacement tube housing assemblies that are reinstalled in or newly assembled into an existing X-ray system are not required to report, as well as the installation of certified, new or repaired, components.

[6]Certified component is defined by the FDA as an electronic product, be it a complete system or a component, that has been certified by its manufacturer that it has been tested and conforms to all applicable standards under 21 CFR sub-chapter J—Radiological Health.

2.1.4 REGULATIONS ENFORCED BY OSHA

Infection Control

Like radiation protection, infection control has the dual purpose of protecting the patients and the workers. In spite of great progress made in microbiology, antibiotics development, and hygiene practices, nosocomial infections are still a major challenge in hospitals. According to Weinstein [1998], approximately 5–6 patients of 100 admitted contract infections in the hospital, and about 88,000 deaths per year are related to the infections. It is, therefore, not surprising that almost all licensing regulations and accreditation standards have extensive requirements dedicated to infection control. Since CE staff is required to visit numerous areas of the hospital occupied by patients, they need to be trained and monitored to follow the hospital's infection-control policies and procedures, including current CDC hand-hygiene guidelines [CDC, 2012], in order to minimize the risk of cross-contamination to patients and clinicians.

In addition, in order to protect themselves and comply with OSHA requirements (in particular, the bloodborne pathogens rules issued by OSHA per 29 CFR 1910.1030), CE staff needs to follow the exposure control plan (including training, universal precautions, engineering and safe work practices, personal protective equipment usage, and post-exposure evaluation and follow-up) established by the healthcare organization where they work.

Material Safety and Hazard Communication

Another mandate from OSHA that hospitals must comply with is the protection against chemicals, known as the hazard communication standard (29 CFR 1910) and enacted in many states as the Employee Right-to-Know regulation. In addition to the chemicals used through the hospital, CE staff is also exposed to special chemicals needed to perform their maintenance duties. Thus, the CE Department is required to keep a material safety and hazard communication program supplemental to that of the hospital for the chemicals that are solely used by CE staff. For these chemicals, the CE Department must provide training, keep chemical inventory and respective material safety data sheets (MSDS), consider replacement or elimination of hazardous chemicals, and collect and provide information to its staff through labels and postings.

Emergency Preparedness Planning and Management

Major natural and manmade disasters in the last decade have highlighted the inadequacies in the planning and management of disasters in hospitals. Again, there are two main perspectives and, thus, different sets of regulations and standards to be observed. The protection of existing patients and provision of care for disaster victims are the focus of the Federal Emergency Management Agency (FEMA) of the Department of Homeland Security (DHS) and similar state, county, and municipal agencies, as well as of licensing and accreditation organizations. The protection of workers is regulated by OSHA through 29 CFR 1910.138.

Since some medical equipment is essential for the delivery of care, the CE Department needs to participate in the planning and management of disasters to ensure that it has appropriate resources to support clinical departments during the crisis. Internally, CE needs to plan for accumulating resources, including alternative backups, for supporting critical pieces of equipment, especially those included in the *DHS Authorized Equipment List* (AEL) issued by FEMA[7]. This plan should include how to secure the following types of resources in case of a disaster:

1. Qualified labor: people who are competent to service medical equipment
2. Parts needed for repairs and scheduled maintenance
3. Tools and equipment
4. Communication equipment

After the plan is completed and harmonized with that for the entire hospital, CE staff needs to participate in the drills and other preparedness tests in order to evaluate and improve the plan.

2.2 STATE LAWS, REGULATIONS, AND CODES

Almost all states have specific laws or codes for licensing healthcare providers, with the purpose of safeguarding public safety and well being. For example, the state of Indiana has Article 16.2—Health Facilities; Licensing and Operational Standards, which is enforced by its State Department of Health, while the state of North Carolina has NC 10A Subchapter 13—Licensing of Hospitals, which is enforced by its Department of Health and Human Services.

The state agencies that enforce licensing regulations and codes often are the same responsible for inspecting the hospitals for compliance with federal CoPs, as part of the state's obligation for receiving and dispersing Medicaid funds. CMS requires states to perform two types of inspections:

1. "validation" surveys—conducted on a small fraction of hospitals accredited by one of the AOs to validate the AOs have performed their surveys properly;
2. "for cause" surveys—conducted as a reaction to complaints filed by patients or their families or reported in the media.

The state inspections are performed following the guidelines provided by CMS in its State Operations Manual (SOM), Appendix A—Survey Protocol, Regulations, and Interpretative Guidelines for Hospitals [CMS, 2012b]. Although this appendix is called "guidelines" for interpreting the code of federal regulations (CFR), its contents have often been considered by many state inspectors as literal requirements. Furthermore, many states have incorporated the CMS Conditions of Participation for Hospitals (42 CFR 482) and SOM guidelines into their own hospital licensing requirements. On the other hand, some states (e.g., Indiana) have issued "waivers" for hospitals to exclude certain pieces of medical equipment from this requirement based on accumulated evidence showing little or no benefit of such inspections, while others (e.g., New York) simply accept as evidence of compliance with its standards successful accreditation with AOA or TJC (10 NY 720.1).

[7] Available at `http://www.fema.gov/library/viewRecord.do?fromSearch=fromsearch&id=5538`.

In addition to licensing, some states have a law that controls the construction and expansion of healthcare facilities, as well as the purchase of sophisticated equipment, known as Certificate of Need (CON). Many CON laws or programs initially were put into effect across the nation as part of the federal Health Planning Resources Development Act of 1974 (HPRD). Despite numerous changes in the past 30 years, including the repeal of the HPRD in 1987, about 36 states retain some type of CON program, law, or agency as of December 2011 [NCSL, 2012].

Almost every state also has its own programs to protect workers and the public. Worker protection is under the jurisdiction of the state occupational safety agency, which typically have laws or codes equivalent to OSHA regulations. Radiation protection requirements vary among states, some with stringent requirements, whereas others mimic federal statutes. For example, Maryland requires registration of all who service ionizing radiation producing equipment, performance of maintenance per manufacturers' recommendations, and submission of preventive maintenance records to its Department of the Environment's Radiation Machines Division (COMAR 26.12.01.01F.3(d)(4)). Most states only require registration of new installations, radiation survey of the facility, and radiation monitoring of workers with routine exposure.

2.3 STANDARDS AND RECOMMENDED PRACTICES

As the name implies, standards are produced by stakeholders (suppliers, producers, users, government agencies, etc.) to ensure consistency of terminology, specifications, performance, and other desired characteristics of products or services. As such, the adoption of standards is voluntary but if adopted by government agency, it becomes mandatory. Recommended practices are even weaker, as it only provides suggested actions. The amount of standards and recommended practices is too large for detailed examination of each one of them. Only a few of them are mentioned below because they are either adopted by at least one federal agency or some state agencies, or are specifically devoted to CE management.

It should be clarified that while the term standards are used by AOs (e.g., the National Integrated Accreditation for Healthcare Organizations—NIAHO standards issued by DNV Healthcare, and the Hospital Accreditation Standards issued by TJC), these documents are mandatory for hospitals that wish to use the accreditation to claim Medicare and Medicaid reimbursements.

2.3.1 NATIONAL FIRE PROTECTION ASSOCIATION

The National Fire Protection Association (NFPA) is a nonprofit organization devoted to reducing fire and other hazards through development of standards, research, training, and education. One of its influential standards is the *NFPA 99—Standards for Health Care Facilities* [NFPA, 2012]. Although it is a voluntary standard, NFPA 99 has been adopted by CMS, many states, counties, and municipalities through reference in their licensing requirements, thus becoming mandatory requirements for hospitals within their respective jurisdictions (but not necessarily for the manufacturers and other organizations located outside).

NFPA 99 covers a wide range of systems within healthcare, such as electrical, gas and vacuum, environmental, materials, electrical equipment, gas equipment, etc., and numerous kinds of hazards, such as fire and explosion, burns, chemical, radio-frequency interference, etc. The section on electrical equipment specifies certain design and construction criteria, as well as periodic tests and inspections. In addition, it contains requirements on user and service manuals, service documentation, and qualification and training of users and maintainers.

2.3.2 ASSOCIATION FOR THE ADVANCEMENT OF MEDICAL INSTRUMENTATION

The Association for the Advancement of Medical Instrumentation (AAMI) is a nonprofit organization with the mission of supporting the healthcare community in the development, management, and use of safe and effective medical technology. Many of the standards and recommended practices produced by AAMI have been approved by the American National Standards Institute (ANSI) as American National Standards.

One of the relevant documents published by AAMI is the Recommended Practice for a Medical Equipment Management Program (ANSI/AAMI [1999] EQ56:1999). This document specifies the minimum required characteristics for a management program designed to minimize certain risks associated with equipment that is used during routine care of patients in a healthcare organization. It provides recommendations for the structure of the program, the documentation that must be produced, and the staffing and resources allocated to those responsible for maintaining the medical equipment. Since it is only a recommended practice, it does not have the authority of a standard and has not been widely adopted by healthcare organizations or government agencies.

CHAPTER 3
Core Functions of Medical Equipment Maintenance and Management

Figure 3.1 shows the primary responsibilities of the CE Department (white blocks) within the lifecycle of a piece of equipment within a hospital. Each of the main CE activities is described in more detail below.

3.1 EQUIPMENT INCORPORATION

Medical equipment often is the single largest capital investment of every healthcare organization, second only to the real estate (i.e., land, buildings, and facility equipment). In addition to high initial investment, the fact that medical equipment requires continual and costly maintenance, as well as supplies and specialized users, means that the initial purchasing cost of medical equipment actually represents only a small fraction of the *total cost of ownership* (TCO), as shown conceptually in Figure 3.2 (see, e.g., Cheng and Dyro [2004]) .

For this reason, healthcare organizations often establish a multidisciplinary Capital Equipment Planning Committee that assists the hospital board and senior executives in their decisions. Figure 3.3 shows an example of how such a committee may be structured. The active participation of the CE Department in every stage of the equipment planning, selection, and acquisition process, collectively known as technology incorporation is highly recommended. As this subject is discussed in detail in a separate book [Wang, 2009], only a brief overview is presented here.

Figure 3.4 shows the three main stages of a medical equipment incorporation process. In the PLANNING stage, a technology audit is conducted to evaluate the existing equipment in terms of functional status, and ability to sustain the operational demand and meet clinical standards of care[8]. Next, an assessment of the needs is determined by the gap between what is ideally needed to fulfill the organization's mission and vision and what the audit reveals is truly available. In parallel, an assessment of the benefits of acquiring new or replacement equipment is made. Using the needs determined, an evaluation of the various kinds of impact (on facilities, human resources, supply chain, etc.) is performed. Finally, a gross estimate of the financial impact and return on investment

[8]Such an audit should actually be performed periodically even if the capital planning were not, as the evolution of healthcare technology is accelerating and some equipment may deteriorate faster than expected.

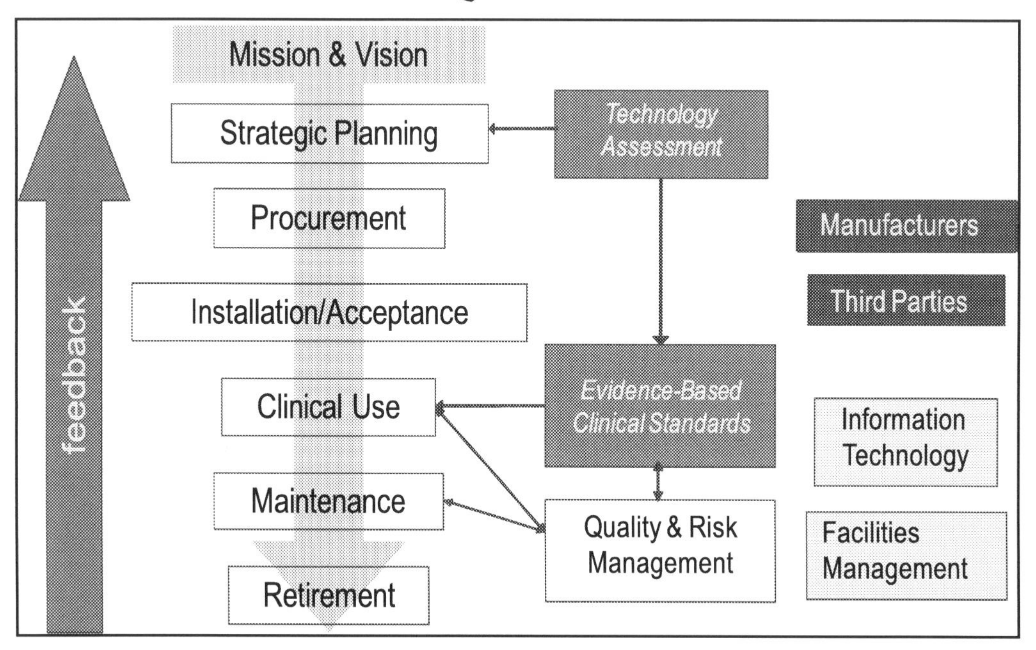

Figure 3.1: Core functions of the Clinical Engineering (CE) department. The primary responsibilities are shown in white boxes following the typical lifecycle of a piece of equipment within the hospital. The organization's mission and vision guide the capital planning process, with input from all stakeholders. The procurement process includes not only purchasing but also other forms of acquisition such as leasing, rental, and consumable-purchase agreements. Technical training and education support is typically provided by CE to users for new and sophisticated equipment. Maintenance, both scheduled and unscheduled (corrective), is the main support activity. Retirement of obsolete equipment, including proper disposal or trade-in, and related replacement is essential to ensure care safety, quality, and productivity. Often neglected but most important is the feedback of actual experience at various stages from all stakeholders so the medical equipment maintenance and management can continually evolve to benefit the organization and its patients.

analysis must be carried out to help decide whether certain equipment requests should be rejected or postponed.

In the second stage (SELECTION), alternative and competing technologies that can address the needs are evaluated, so the hospital can reduce the risk of acquiring equipment that is likely to be phased out soon due to the emergence of better or newer solutions. Similarly, products that are potential alternatives to the ones initially considered for the needs identified should now be searched and studied in detail. Both studies should consider where each technology or product is located in its product lifecycle from the perspective of the user and not from the producer, as the latter

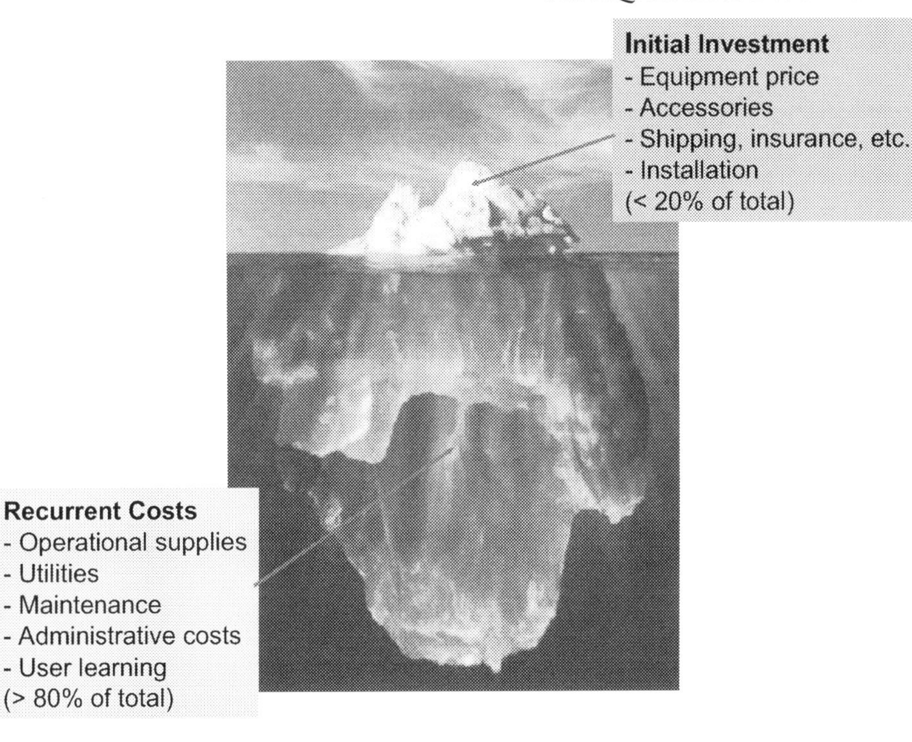

Figure 3.2: Graphical representation of the main components of the total cost of ownership (TCO) of medical equipment within a hospital. Typically, the initial investment, including purchase price of the equipment and accessories, shipping, insurance, installation, licensing, training, etc., add up to less than 20% of the TCO. The remaining initially "invisible" expenses are incurred during the useful life of the equipment (sometimes as long as >15 years), in the form of supplies, utilities, maintenance services, upgrades and updates, administrative overhead, user training, self learning, etc.

tend to emphasize investments that provide high profit to them but also risk to the user (for more details, see Wang [2009]). Based on these analyses, the assessments of benefits and impacts previously conducted may have to be updated or even revised extensively. Similarly, the initial financial estimates will now need to be reviewed and calculated with the new data available.

In the third and final stage (ACQUISITION), the desired equipment will be acquired and processed so it can be used for patient care. The acquisition process should not be limited to the traditional purchase or even lease, but also rental, loan, supply-purchase (also known as "reagent-rental") agreement, and donation. Some capital equipment will require installation with specific demands on utilities, radiation protection, etc., so a diligent and close working relationship with the vendor is essential to ensure successful installation. In the acquisition process, the CE Department can provide valuable input with regard to warranty, post-warranty service, user technical training,

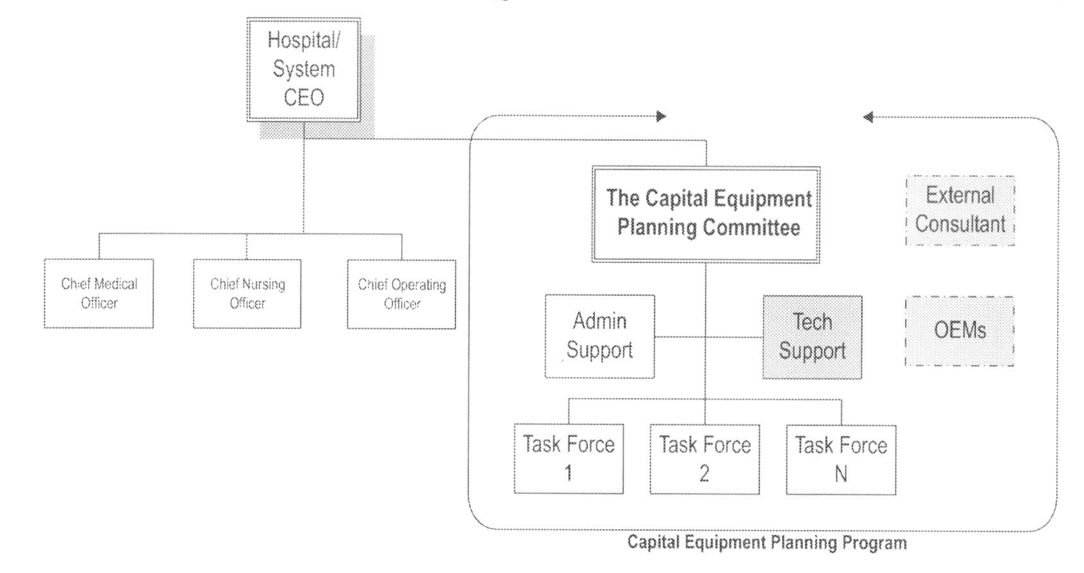

Figure 3.3: Example of the relationship of the Capital Equipment Planning Committee within a hospital structure. The CE Department typically provides the technical support (green block) for all medical equipment related studies, both at the level of the entire committee and for individual task forces that are focused on individual technologies. In addition to medical equipment issues, CE staff is often involved in discussions involving other capital equipment that have direct or indirect impact on medical equipment. This input may extend to all equipment due to their technical expertise and analytical skills.

service documentation, maintenance training, availability and cost of supplies and parts, software upgrades, etc., to minimize the TCO and maximize equipment availability.

In all three stages of the equipment incorporation process, the CE Department should participate intensely. Some hospitals have even adopted the practice of requiring the signature of the CE Department manager before any purchase order for capital equipment can be issued. To ensure the best discount available, along with assuring all technical issues are addressed (compatibility, regulatory requirements, utilities, etc.) some purchasing departments defer to the CE Department for coordination of quotes. This offers the additional benefit of streamlining the purchase process. In fulfilling these duties, the CE Department will use its internal expertise and experience, as well as technology-assessment information provided by independent and reputable organizations, such as the Advisory Board, ECRI Institute, and MD Buyline.

Obviously, not every medical equipment incorporation needs such an extensive and thorough process. Most hospitals have created one or two thresholds in terms of total acquisition costs to decide the level of scrutiny needed. However, large amounts of mid-level devices (e.g., 500 infusion pumps) deserve the same attention as a single piece of high-cost equipment due to the TCO involved.

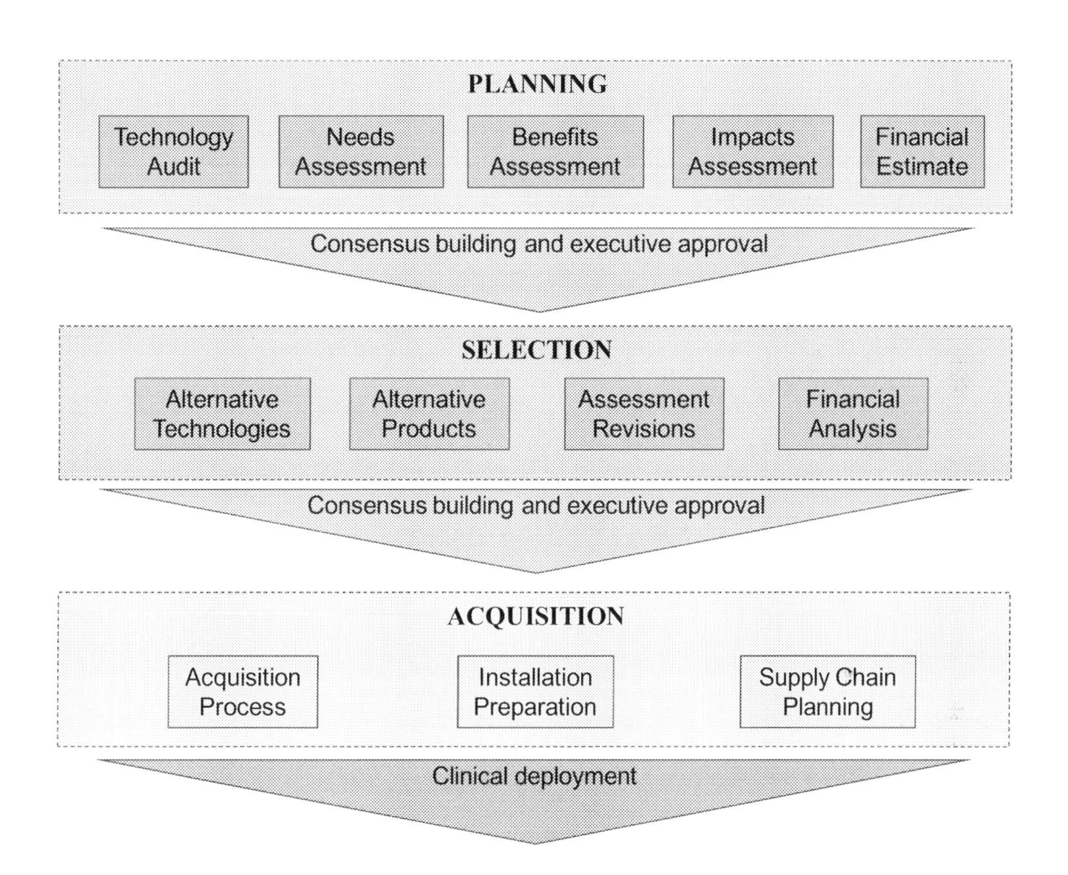

Figure 3.4: Medical equipment incorporation process. The planning stage includes technology audit, needs assessment, benefits assessment, impact assessment, and a preliminary financial estimate of the costs and return on investment. The consensus achieved by the Capital Equipment Planning Committee is reviewed and approved by the appropriate authorities before proceeding to the next stage. The selection stage involves evaluating alternative or competing technologies, searching for alternative products, revising the assessments made earlier if needed, and a more detailed and comprehensive financial analysis. Again, after consensus building and approval by the appropriate executives, the process proceeds to the next stage. The acquisition stage includes not only the *sensu stricto* acquisition but also all the related activities of installation preparation and planning for supplies purchase and storage. When all the three stages are followed carefully, the success of the clinical deployment is greatly enhanced.

3.2 INSTALLATION AND ACCEPTANCE

Some newly acquired equipment needs installation (i.e., affixed to the building and connections to water and gas lines), while others can be used right out of the box. In either case, a series of qualitative and quantitative tasks designed to verify the safety and performance of the new equipment must be performed per CMS guidelines and accreditation standards. Furthermore, verification of conformity to applicable codes, regulations, and standards (e.g., tested or listed by a Nationally Recognized Testing Laboratory (NRTL) recognized by OSHA, such as the Underwriter Laboratories (UL) and the Canadian Standards Association (CSA)) must be conducted. As explained before (see Section 2.3.1), depending on the status of adoption of NFPA 99 standards by the state agency having jurisdiction, compliance of the equipment to this standard also needs to be verified.

Finally, acceptance should only be granted if the equipment is delivered with the operating and service documentation, and the provision of user and service training, as required in the purchase order. Payment should not be authorized until the receipt of all the promised accessories, technical documentation, and training (both clinical and for the CE staff) is verified.

3.3 USER TRAINING

Root-cause analyses performed by healthcare organizations and review by TJC show the primary cause of medical equipment-related patient incidents is human factors, i.e., misuse caused by unintuitive design and instructions [TJC, 2012a]. A classical example is the so called "alarm fatigue" [AAMI, 2011a]. Clinicians are often overwhelmed by the amount and diversity of alarms generated by medical equipment, so they simply ignore the alarms altogether. This obviously becomes unsafe when a truly serious alarm occurs and the patient does not receive prompt attention.

Before a new piece of equipment is deployed in any healthcare organization, the clinical users should receive appropriate in-service training from the respective manufacturer, including proper clinical indications, potential risks and side effects, pre-use operational checks, and basic equipment functions and troubleshooting. This is particularly important when the new equipment is designed differently from the one it replaces and when the hospital relies on agency nurses. CE staff should participate in these in-service trainings, as well as technical training for servicing the equipment. These trainings will enable CE staff to provide clinical users with technical (but not clinical) refreshers later whenever needed.

3.4 EQUIPMENT MAINTENANCE

This is the activity that most healthcare professionals remember when CE departments are mentioned. While it is still the "bread and butter" of what CE professionals do, it is not the most important function nor does it have the highest impact. As explained in the remainder of this book, maintenance is only part of the overall medical equipment management. Without a proper incorporation process (i.e., acquiring the right product in the right manner at the right time), inappropriate equipment could be acquired that does not produce effective and efficient patient care. Without a continual, systematic retirement and replacement process (i.e., replacing an undesirable product in the right manner at the right time), unreliable, obsolete, or unsafe equipment could be used beyond its useful life, putting patients and users at risk. Most importantly, without proper consultative assistance from CE professionals, healthcare organizations are likely to waste precious capital resources, increase operational expenditures, reduce patient throughput, and increase unnecessarily risks to users and patients.

Two other misunderstandings need to be dispelled before addressing how maintenance should be planned, implemented, monitored, and improved continually. The first one is the presumption that "maintenance guarantees patient safety" or "the purpose of maintenance is to ensure patient safety." Although the CE discipline gained much impetus from the headlines of electrocutions published in the early 1970's [Nader, 1971][9], over three decades of work have proved that the main goal of CE is not patient safety in *sensu stricto* but reliability and, consequently, availability of equipment for patient care. Obviously, one could argue that increased equipment reliability translates into patient safety in *sensu latus*, as it helps to ensure that patients get prompt diagnosis and treatment that nowadays depend strongly on mission-critical devices such as automated laboratory chemistry analyzers, CT scanners, and remote patient monitoring systems. Like the aircrafts used in commercial aviation, maintenance by itself cannot guarantee passenger safety, but it is undeniably an important link in the safety process [Moubray, 1997; Smith and Hinchcliffe, 2004]. Furthermore, passenger safety is not the only goal of aircraft maintenance. Airlines cannot survive if the passengers are safe but do not arrive at their desired destinations without excessive delays and inconveniences, and at competitive prices.

The other misunderstanding that has persisted for many years in almost all industries is the belief that "the more maintenance the better." Data collected from numerous industries has proven that this is not the case (see, e.g., Corio and Costantini [1989]; Ackoff and Rovin [2005]; Zak [2005]). Actually, like clinicians, CE staff could also commit errors that could have significant or even catastrophic consequences on patients and clinical users. The only scientific way to evaluate the effectiveness of maintenance activities is to assess the outcomes in terms of equipment reliability, much the same way clinicians have been evaluating clinical outcomes over the last couple of decades, a practice known as *evidence-based medicine* [EBMWG, 1992]. The analogous concept would be *evidence-based maintenance* [Wang, 2007, 2008; Wang et al., 2010a,b, 2011], which will be explained better in the Discussion.

Detailed analyses of patient incidents collected by the FDA [Bruley, 1998] and TJC [Wang et al., in press] show that the amount of equipment related incidents that can be traced to maintenance omissions (i.e., lack of or improper maintenance) is less than .00011–.0006 per million equipment use, which is significantly lower than the widely known Six Sigma criteria of 3.4 defects per million. This suggests that medical equipment maintenance has been performed and managed exceedingly well, with perhaps excessive resources invested than what is actually needed due to the lack of evidence-based studies, as well as liability and negative publicity concerns.

[9]Subsequent research has shown that those allegations were unfounded [Ridgway et al., 2004].

With these misunderstandings clarified, one can now explain how medical equipment maintenance should be planned, implemented, monitored, and improved continually.

3.4.1 MAINTENANCE PLANNING

The main challenge for the CE Department in maintenance is to find the appropriate strategy to care for each piece of equipment. As mentioned in the beginning, each hospital has typically around 15-20 pieces of medical equipment for each staffed bed, so it is common for a 500-bed hospital to have over 7,500-10,000 pieces of equipment, without considering items that cost less than $1,000 each such as surgical instruments, stethoscopes, clinical thermometers, etc. This large quantity is further compounded by multiple and, sometimes, inconsistent regulatory requirements issued by different agencies (e.g., CMS, FDA, TJC, AOA, and state departments of health and licensing agencies) and voluntary standards issued by various organizations (e.g., American Association of Blood Banks—AABB, College of American Pathologists—CAP, and National Fire Protection Agency—NFPA).

Figure 3.5 shows graphically the main elements that need to be considered in maintenance planning. The outmost oval represents the entire universe of medical devices deployed in a hospital, including not only capital assets but also single and multiple-use devices. The next oval represents the assets that are inventoried because their individual acquisition cost exceeds a certain threshold (e.g., $1,000) and are "durable" (i.e., not an implant, single-use or single-patient use device). Within this oval, the diversity of devices (ranging from a laboratory microscope to an MRI) is represented by the gradient in gray shading. The equipment that falls into this oval is what each CE Department must analyze and decide which maintenance strategies should be deployed. The three main elements for maintenance decision are:

1. *patient safety*: the risk of an equipment failure in harming a patient, i.e., the combination of the probability of harm and the severity of harm (ISO/IEC Guide 51:1999, ISO 14971, 2000);

2. *intrinsic maintenance needs*: maintenance actions required by the equipment due to its design and the components used for its construction, i.e., the combination of preventive maintenance (replacement of wearable parts, lubrication of moving elements, etc.) and safety and performance inspections (detection of hidden and potential failures, as defined below);

3. *criticality to the organization's mission*: the likelihood and severity of the impact of an equipment failure on the hospital's mission of providing care to patients, including the effects of delayed and/or interrupted diagnostics and therapeutics, and also the consequential impacts on patient satisfaction and institutional finances [Wang et al., 2006a].

As shown in Figure 3.5, these three elements are not always present in every piece of equipment and even when they are, not with the same intensity. In other words, equipment that is maintenance intensive is not necessarily critical for patient safety and *vice versa*. Therefore, a process of prioritizing the attention (i.e., effort of CE staff) is needed. Over the last three decades, several methods of prioritization have been used [ASHE, 1982, 1996; Fennigkoh and Smith, 1989; Wang and Levenson,

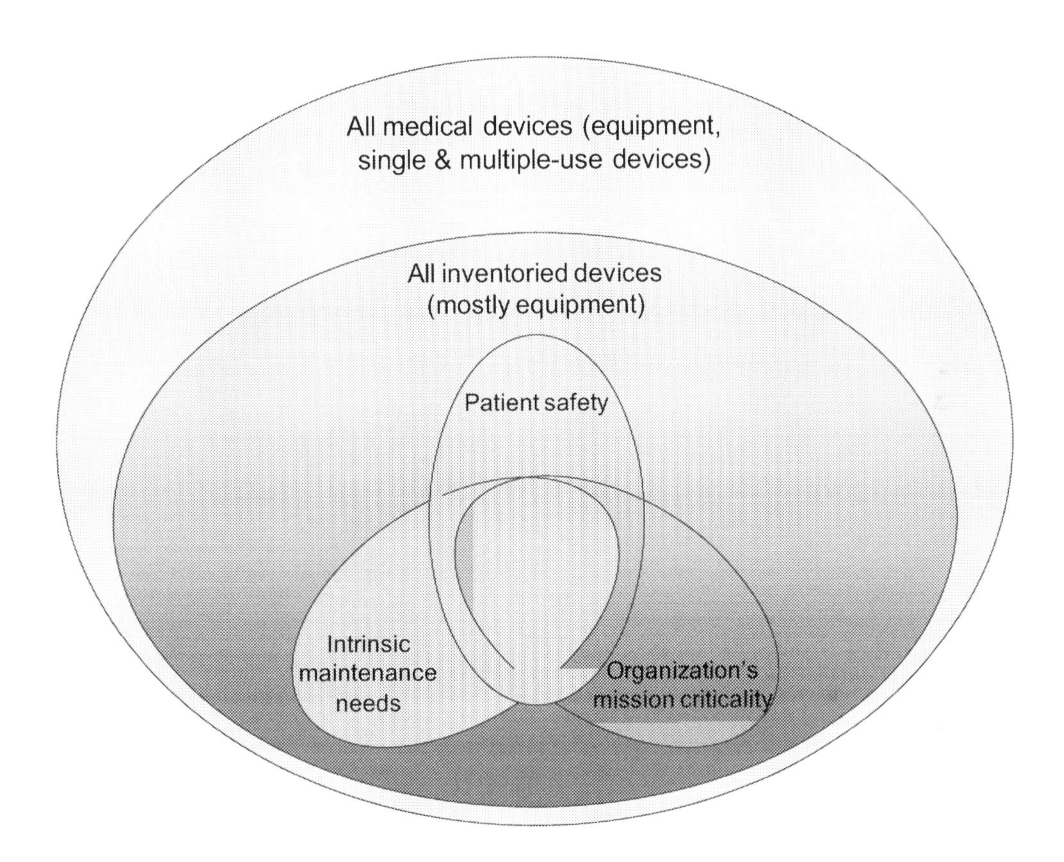

Figure 3.5: A graphical representation of the elements that need to be considered by the CE Department in determining the best maintenance strategy for each piece of equipment. The outermost oval represents the entire universe of medical devices, including implants, single and multiple-use devices, that exists in a hospital. The inner oval represents all inventoried assets (aka "capital devices") that typically have purchase cost above a certain threshold (e.g., US$1,000). The shading of the inner oval calls attention to the fact that it is not a uniform set of devices. It is rather a continuum of a wide variety of devices with distinct complexity, risk, cost, and maintenance requirements. The three smaller ovals represent equipment that present reasons to be considered in prioritizing service attention: patient safety, criticality to the organization's mission (delivery of care), and inherent maintenance needs imposed by the equipment internal components. Notice that these three smaller ovals do not overlap completely, as each one exists independently of the others. The challenge to the CE Department is how to attribute priority starting from the areas with overlaps and moving out gradually to those that have less or no overlap.

2000; Wang et al., 2006a; ECRI, 2012]. As a full description and discussion of this subject would require too much space, readers are referred to Wang [2008] for an in-depth treatment. Suffice to state here that using the three elements described above, the CE Department should create a service strategy for each piece of medical equipment that is suited to the genuine needs of each hospital.

3.4.2 MAINTENANCE IMPLEMENTATION

After appropriate maintenance strategies have been determined, their implementation requires the following types of resources:

1. Human resources: staff with appropriate technical competency and customer-service skills in enough quantity to perform the work planned;
2. Material resources: physical space, workbenches, test and measurement equipment, tools, and other instruments necessary to perform the inspections, preventive maintenance, and repairs;
3. Financial resources: money to acquire the maintenance and replacement parts, subassemblies, etc., needed to maintain and restore the equipment functions; and
4. Documentation: operating and service manuals, software programs, parts lists, electronic/electronic and mechanical drawings, etc.

Unfortunately, none of the first three resources can be determined using a simple mathematical formula. Each hospital has apparently found a slightly different way to fund and operate its CE Department to achieve a level of service that it finds satisfactory, often blending work performed by onsite staff with assistance from manufacturers and alternative sources (see below for a more detailed discussion about service provider selection). Nonetheless, a ballpark estimate for the necessary human and financial resources is possible using a multi-dimensional model [Wang et al., 2012].

3.4.3 MAINTENANCE MONITORING

For many years, the only kind of monitoring conducted by CE departments and hospital administrators have been the completion of scheduled maintenance services, because this is the requirement of regulatory and accreditation organizations. For example, TCJ standards [TJC, 2012b] require one-hundred percentage completion of scheduled maintenance (SM) on life support equipment. While TJC does not specify a completion target for non-life support equipment it reserves the right to issue a "requirement for improvement" (RFI) if its surveyor finds three or more pieces of equipment that are overdue for scheduled maintenance (even when a hospital or system has over 10,000 pieces of equipment scheduled for maintenance).

The singular emphasis on a scheduled maintenance completion rate reflects the misunderstanding of "maintenance guarantees patient safety" discussed above. Just as educators have long understood that classroom attendance does not guarantee student learning, CE managers need to shift their attention from completion rates to outcomes measures. From the clinical users' perspective, the most important outcome of a good maintenance program—actually from the entire CE program, including all aspects of equipment incorporation and management—is reliability, which is usually measured by the availability of equipment for clinical use. For mission-critical, heavily used equipment (e.g., MRI, CT, linear accelerator, etc.), availability is typically measured by its uptime, defined as the percentage of time the equipment is available for use as compared to the total time it is planned and staffed for use[10]. For non-mission-critical equipment (e.g., infusion pumps, patient monitors, etc.), uptime is difficult to measure—and, perhaps, even meaningless—because normally there are enough backups available to replace the failed piece so patient care can continue without interruption. In these cases, it is common to use failure rate, defined as the number of verified[11] failures per year as a percentage of the total number of pieces of equipment in the inventory. Failure rates should be calculated ideally for each brand and model of equipment, but it is sufficient in most cases to aggregate different brands and models that have the same function and similar principle of operation (e.g., all volumetric infusion pumps instead of only Brand A, model 1000). A global failure rate for a wide class of equipment (e.g., imaging, laboratory, and biomedical) can also be useful [Wang et al., 2006a] but it does not provide enough detail for determining opportunities for improvement.

In addition to monitoring the outcome indicators, CE managers need to understand better the failures detected in the medical equipment and their respective causes and potential solutions. Until recently, most CE departments have attempted to record every part failure and corrective action taken. While such records can be very useful for the manufacturers to determine individual component reliability or the CE Department to stock parts, it does not help CE managers to understand what can be done in terms of maintenance strategy to enhance reliability. Instead of recording only the parts replaced or corrective and preventive actions taken, CE staff should be trained to conduct a rudimentary root-cause analysis of each SM and repair performed and assign a code to each service record. When a large number of records have been collected, the CE manager can determine where the opportunities for improvement are. One such set of failure cause codes is shown in Table 3.1 [Wang et al., 2010a,b, 2011]. The codes that offer the best opportunities for improvement are:

1. Preventable and predictable failure (PPF): a failure that is typically caused by wear and tear that can be predicted or detected. The appropriate solution is scheduled maintenance for detecting—*safety and performance inspection*—and/or replacing the worn part(s)—*preventive maintenance*.

2. Potential failure (PF): a failure that is either about to occur or in the process of occurring but has not yet caused equipment to stop working or problems to patients or users. The appropriate solution is scheduled inspection that would determine the need for replacement of defective or worn part(s) or calibration adjustments.

[10] There are numerous variations of the definition of uptime. Some would exclude after hours even though clinical services are routinely scheduled beyond business hours. Others would limit failures only to those that prevent the equipment to be used at all, while partially functioning equipment would not be considered "down."

[11] Verified failures are those that were duplicated by CE staff and not misunderstandings by the user on how the equipment should perform, commonly termed "cannot duplicate."

Table 3.1: An example of failure cause codes that can be used to help determine appropriate adjustments to the maintenance program. The first group includes equipment failures while the second shows peripheral failures (accessories, supplies, or network). Within the first group, failures are further divided into those found during unscheduled, corrective maintenance (repairs) and those found during scheduled maintenance (inspections, calibration, and preventive maintenance).

SOURCE	ACTIVITY	CODE	CODE DEFINITION
Equipment	corrective maintenance (CM)	**UPF**	Unpreventable failure, evident to user, typically caused by normal wear and tear but is unpredictable.
		PPF	Preventable and predictable failure, evident to user, typically caused by wear and tear that can be predicted or detected.
		USE	Failures induced by use, e.g., abuse, abnormal wear & tear, accident, or environment issues.
		SIF	Service-induced failure, i.e., failure induced by corrective or scheduled maintenance that was not properly completed or a part that was replaced and had premature failure ("infant mortality").
	scheduled maintenance (SM)	**PF**	Potential failure, i.e., failure is either about to occur or in the process of occurring but has not yet caused equipment to stop working or problems to patients or users.
		EF	Evident failure, i.e., a problem that can be detected but was not reported by the user without running any special tests or using specialized test/measurement equipment.
		HF	Hidden failure, i.e., a problem that could not be detected by the user unless running a special test or using specialized test/measurement equipment.
	CM and SM	**NPF**	No problem found, including alleged failures that could not be duplicated ("cannot duplicate"– CND)
Peripherals or Network	CM or SM	**BATT**	Battery failure, i.e., battery(ies) failed before the scheduled replacement time.
		ACC	Other accessory failures, excluding batteries, evident to user, typically caused by normal wear and tear.
		NET	Failure in or caused by network, while the equipment itself is working without problems. Applicable only to networked equipment.

3. Hidden failure (HF): a problem that could not be detected by the user unless running a special test or using specialized test/measurement equipment. The appropriate solution is scheduled inspection that would attempt to detect the failure and, if found, correct the failure.

3.4.4 MAINTENANCE IMPROVEMENT

Using maintenance monitoring data collected, CE managers can find ways to improve maintenance services. For example, if the maintenance outcomes (uptime or failure rate) are unsatisfactory even though the scheduled maintenance completion rates are high, the amount of maintenance opportunities detected (PPF, PF, and HF) need to be reviewed and their respective causes uncovered. On the other hand, if PPF, PF, and HF are seldom found during scheduled and corrective maintenance, the frequency of scheduled services may be unnecessarily high.

Frequent reviews and adjustments of individual maintenance strategies are typically necessary when CE staff is not familiar with a particular brand or model of equipment. After desired outcomes are achieved, it is not necessary to revisit every equipment group unless some unforeseen problems appear. One way to optimize resources would be to share experiences with CE departments in similar healthcare organizations that deploy the same brands and models of equipment. This kind of cooperation is common in service outsourcing organizations where they can easily pool data and experience from dozens or even hundreds of hospitals. Otherwise, one has to rely on the goodwill of fellow CE professionals. Unfortunately, manufacturers are often reluctant to assist in the optimization of maintenance efforts. Apparently, some are concerned about their liability exposure if they become too aggressive in their recommendations, while others see post-sale service as a major source of revenue.

The discussions above are limited to improvement of maintenance activities. A broader discussion of CE Department performance monitoring and continual improvement will be presented later in this book.

3.5 EQUIPMENT REPLACEMENT, RETIREMENT, AND DISPOSAL

Like any other type of equipment, medical equipment needs to be replaced when it no longer provides safe, reliable, and effective services. Due to the rapid introduction of new and better medical technologies and procedures, medical equipment is increasingly being replaced well before it is unsafe or unreliable. Nonetheless, the CE Department has the duty to review periodically the status of each piece or group of equipment to determine whether it is no longer technically feasible or financially justifiable to continue to maintain it. Once replacement seems needed, the CE Department should recommend the Capital Equipment Planning Committee to consider it as part of the equipment incorporation process discussed above. Again, a detailed discussion on how to evaluate the technical and financial aspects of equipment replacement is beyond the space allotted for this book but can be found in a prior one [Wang, 2008].

Once the decision of replacing a piece of equipment is reached, the CE Department should assist in the retirement and disposal of the replaced unit. Several possibilities should be considered, such as keeping it as a backup, transferring to another facility, trading in for the new one, selling it to another hospital or a reseller, donating it to an appropriate recipient in or outside of the country, and taking it to a recycling center. CE department involvement in each of these options is essential to ensure that the process is successful, and relevant environmental regulations, codes, or standards (e.g., disposal of hazardous materials) are observed.

3.6 SAFETY AND RISK MANAGEMENT

Patient safety is an inherent duty of all healthcare professionals. Since absolute safety is all but impossible, especially in healthcare, a systematic method to identify, evaluate, reduce, or eliminate unfavorable outcomes is essential. Even when all these steps have been performed correctly, undesirable results can still happen. For this reason, hospitals safeguard themselves with insurance and set aside funds to cover claims. The person responsible for all this in the hospital is the risk manager.

Obviously, the risk manager cannot know every aspect of healthcare in detail, so he/she needs cooperation from each clinical and support department in order to fulfill his/her responsibility. A close working relationship between risk management and CE helps the latter to appreciate legal aspects beyond the laws, regulations, codes, and standards applicable to healthcare and CE. Conversely, CE can help risk management to understand how medical equipment risks can be addressed. Furthermore, whenever there is an incident, they can work together to determine the root causes and opportunities to improve their existing policies, procedures, and processes.

Risk is defined by international standards as the combination of the probability of occurrence of harm and the severity of that harm (ISO/IEC Guide 51, 1999, ISO 14971, 2000), i.e.,

$$\text{Risk} = P * S \tag{3.1}$$

where P is the probability of harm and S is the severity of the associated harm.

The severity of harm is often intrinsic to individual drugs, medical devices, and medical procedures, because each one has some side effects and limitations that cannot be completely eliminated. Therefore, the primary role of risk management is to decrease the probability of harm as much as possible within the available resources. In this context, CE contributes to hospital risk management by reducing the probability of equipment failures using a combination of maintenance (scheduled and unscheduled) and management (incorporation of appropriate equipment, management of recalls, corrections and upgrades, incident investigation, and replacement of unsafe or obsolete equipment). As maintenance, incorporation, and replacement have been covered above, the other two are discussed briefly below.

3.6.1 RECALLS, CORRECTIONS, AND UPGRADES

As explained in Section 2.1.3, recalls are actions taken by a manufacturer or its representative to remove from the market a product that can pose risks to health or violate FDA regulations. To remediate the problem, the manufacturer may remove the product or make corrections through inspections, updates, or replacement of affected parts.

Since recalls are covered by federal regulation (21 CFR 7, 80, 806, and 810), compliance by manufacturers and healthcare organizations is mandatory. The FDA conducts some recall effectiveness verifications with healthcare organizations by visiting recipients of the recalled products and asking for evidence that the recall has been received and the necessary actions have been completed.

3.6.2 INCIDENT INVESTIGATION

As explained in Section 2.1.3, hospitals are required to report serious injuries and deaths related to medical devices to the respective manufacturers and, in the case of deaths, to the FDA.

While AOs require hospitals to comply with FDA regulations (often mentioning SMDA of 1990), they don't require reporting of these incidents. TJC calls these incidents "sentinel events" and encourages the hospitals that it accredits to report them with root cause analyses[12]. TJC also performs statistical analysis on sentinel events to offer insights into their root causes and potential solutions.

Besides the regulatory and accreditation requirements, hospitals need to be well prepared to manage patient incidents in order to reduce liability risks and improve patient safety. A valuable tool in this process is incident investigation because it allows the hospital to determine the root causes of each incident. As explained by Reason with his "Swiss cheese" model [Reason, 2000], incidents rarely have a single cause. For example, if a piece of equipment fails to operate, it will only cause patient harm if the failure is not detected in time by a clinical user (e.g., undetected alarms) and the patient him/herself is unable to react against the malfunction. This can be visualized more easily if one were to rewrite Equation (3.1) into

$$\text{Risk} = \left(\prod_i P_i\right) * S \tag{3.2}$$

where P_i represents the probability of harm for failure i. In other words, P in Equation (3.1) is replaced by the multiplication of individual probabilities of failure, such as the probability of equipment failing to function properly, the probability of a clinical user failing to detect the alarm, and the probability of the patient failing to withstand or react to the equipment malfunction.

Because of potential legal implications, incident investigations must be performed according to a well established protocol led by the hospital's risk manager (or legal department). This typically involves someone who can test the equipment, someone who understands how it works normally, someone who will take accurate notes, and someone who can serve as witness to the process. While some hospitals prefer to hire an independent third-party investigator, this is often not needed if the CE staff has the competency to test the equipment, as long as the equipment is not opened or otherwise tampered with, and a witness is available. Testing by a manufacturer representative is a

[12]For more information, see `http://www.jointcommission.org/sentinel_event.aspx`.

good alternative with similar precautions. However, shipping the equipment to another location for testing should be avoided due to potential damage and loss during shipment and risk of tampering.

CHAPTER 4

CE Department Management

In order to fulfill the functions described above, CE departments must organize themselves and adopt a number of processes to manage themselves so they can deliver its services in the most cost-effective manner. This can be quite challenging for hospitals with limited resources or located in areas where CE talents are difficult to recruit and retain. For this reason, at least 30% of acute care hospitals have opted to outsource such operations to companies that offer these services, so the former can focus on their core clinical functions.

4.1 MANAGEMENT OF PROCESSES AND DOCUMENTS

One of the fundamental responsibilities of the CE manager that is often relegated to lower priority and sometimes even neglected is the documentation of the internal business processes in the form of policies and procedures. As explained above, CE professionals are typically strong in technical expertise and less proficient with administrative duties. Writing and updating policies and procedures continually can be rather challenging and time consuming. However, without proper documentation, the processes for planning, acquisition, maintenance, and management of equipment will simply reside in the memories of staff and, thus, likely lead to inconsistent implementation and unpredictable outcomes.

One approach that can help simplify the documentation management is to adopt a documentation hierarchy inspired by ISO 9001 standard [ISO, 2008] as shown in Figure 4.1. At the top of the pyramid reside the policies, which are statements of principles adopted for management of equipment, staff, and operations. These statements can be as short as a single sentence and at maximum, a short paragraph, and typically can remain unchanged for many years, even when there are changes in regulations, codes, and standards. The next level is composed of the procedures, which describe the processes used to implement the policies. The procedures are more detailed and may need to be revised periodically to reflect changes in the hospital's structure or business practices, as well as changes in regulations, codes, and standards.

The third level is composed of work instructions, i.e., procedures for performing scheduled and/or corrective maintenance on specific brands and models of equipment, including service manuals published by equipment manufacturers. The lowest level is composed of records of work performed, including labels attached to equipment and electronic records.

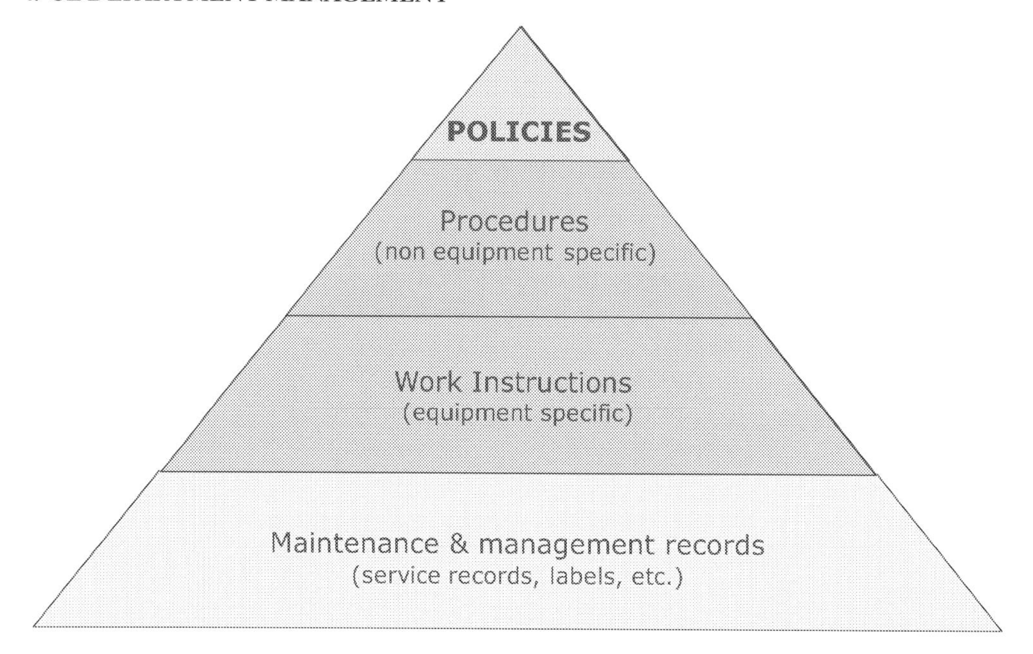

Figure 4.1: Documentation hierarchy. Internal management documents are divided into four levels according to their relative importance and scope. Policies are declarations of principle that must be followed at all times. Procedures describe processes used to implement the policies and are applicable to a wide range of equipment, sometimes even at different locations. Work instructions provide the tasks needed to perform scheduled and unscheduled maintenance and other types of services. Records provide proof of work performed and information for users or other maintenance workers.

4.1.1 POLICIES AND PROCEDURES

As explained above, the policies state the principles adopted by the hospital and the CE Department for the management and maintenance of medical equipment, reviewed and approved by the hospital Safety Officer or Committee (or equivalent authority). An example of policies is shown in Appendix A.

Procedures provide the description of the processes used to implement the principles stated in the policies. Unlike policies that are often limited to a single sentence or paragraph, procedures typically have several pages as they need to provide answers to the classical five questions: who, what, why, when, and how. Therefore, it is common that the CE Department procedures are found in a binder that is over 1 in thick. Nowadays, more often procedures are posted on secured intranet sites where they can be easily updated and found by the staff when needed.

Together, the policies and procedures synthesize the numerous requirements, goals, and objectives that the CE Department has to observe and comply with in order to ensure that it helps the healthcare organization fulfill its mission, vision, and values, while complying with applicable laws, regulations, codes, standards, internal rules, and industry standards and accumulated experience (Figure 4.2).

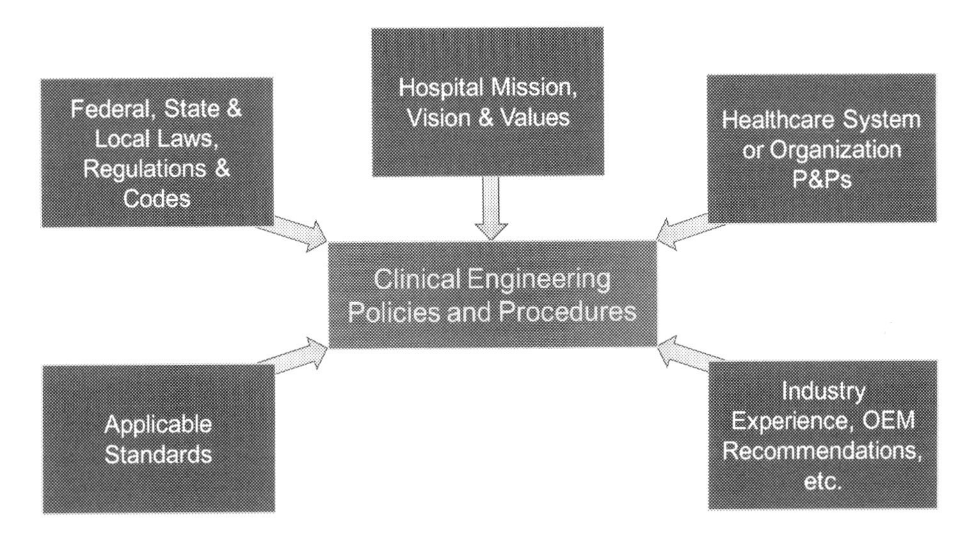

Figure 4.2: A graphical representation of the elements (e.g., hospital mission, vision, and values; laws, regulations, codes, and standards; healthcare system or organization internal requirements; industry experience and manufacturers' recommendations) that are incorporated into the policies and procedures of the Medical Equipment Management Program.

Although not required by regulation or licensing codes, both policies and procedures should be reviewed and approved by the hospital's Safety Committee, Officer, or equivalent authority, as this provides a "check-and-balance" approach instead of a unilateral action from the CE Department. Furthermore, it is highly desirable that the stakeholders (i.e., leaders of the clinical and support departments whose operations depend heavily on or affect medical equipment) participate in the reviews and approvals. The distinction between policies and procedures can help to reduce the amount of work needed for periodical review and revision, as policies seldom need revision while procedures are likely to require modifications to reflect organizational changes, adoption of new technologies and processes, and improvements for gains in efficiency and patient outcomes.

4.1.2 WORK INSTRUCTIONS

Work instructions provide the details needed by CE staff to perform scheduled and corrective maintenance services. While work instructions are sometimes provided by manufacturers and professional organizations [ASHE, 1996; ECRI, 2012], the CE Department often needs to develop its own instructions in order to improve efficiency.

Service Documentation (Manuals)

As stated above, operating and service documentation must be demanded from the respective equipment manufacturer when acquiring equipment[13]. This is because these manuals contain the scheduled and corrective maintenance work instructions, as well as other essential information such as operator's instructions for proper use and care, cleaning and disinfection process, list of accessories and parts, and warranty information. For this reason, these manuals should be carefully controlled by the CE Department to prevent loss and allow prompt retrieval whenever needed. CE Department staff should not be allowed to write or alter, remove pages from, or otherwise damage them.

Service Records

It is critical to ensure proper recordkeeping of the services performed. Documentation of services performed is not only required by regulatory and accreditation organizations but also for legal defense in case of lawsuits involving equipment malfunction. These records can be divided into two broad classes: full records and abbreviated records. The former are comprehensive descriptions of the service performed, including typically the reason for service, scheduled or unscheduled actions, parts replaced, etc. The full records can be on paper records but nowadays most of them are entered and stored with a CMMS which is discussed in Section 4.2 below. The abbreviated records are labels (aka "stickers") attached to the equipment with basic information such as the dates of service completed and next scheduled service, initials of the CE staff who performed the service, and sometimes some of the measurements. The abbreviated records should be considered secondary to the full records (primary) because the labels often become illegible or lost during the frequent disinfection processes.

Retention of records is governed by the hospital's *Record Retention Policy*. After the retention period, records should be destroyed to reduce risk of inadvertently releasing confidential or private patient information that could be found in some of these records, as required by the HIPAA regulation.

4.2 COMPUTERIZED MAINTENANCE MANAGEMENT SYSTEM (CMMS)

A CMMS is typically used by CE departments to manage and maintain equipment inventory, schedule service, record repairs, keep service records, track expenditures, and produce reports. Ideally, a CMMS should be able to be customized to fit each hospital and CE Department's needs in terms of not only complying with regulations, codes, and standards applicable to its environment, but also fulfill users and CE staff needs in terms of easy access to the information and production of reports. Numerous commercial CE CMMS software products are available (see, e.g., Cram [1998]) and can

[13]There is no legal requirement for manufacturers to provide (even for a fee) service manuals to equipment buyers or owners. The FDA only requires manufacturers to provide specifications for assembly, installation, adjustment, and testing (AIAT) for X-ray systems and related major components (21 CFR 1020.30(g)). While NFPA 99 standard does have a clause with such requirements, enforcement has been difficult as most states that have officially adopted this standard do not have the means to pursue violators.

satisfy the needs of most CE departments, so typically it is not justifiable to develop one unless there is a peculiar need and strong commitment for its long-term maintenance.

In addition to the basic features, the CE manager should also consider other aspects when selecting or upgrading the CMMS, such as information security, data backups, and restore capabilities, and training needed for clinical users, administrative staff, and CE staff. Like the incorporation process for medical equipment, the total cost of ownership (TCO) of a CMMS needs to be carefully calculated and analyzed to ensure its cost effectiveness.

4.3 STAFF MANAGEMENT AND DEVELOPMENT

Like other service organizations, the most important and difficult resource to quantify and qualify, as well as to manage, are the human resources. Because of the complexity of the modern medical equipment, the CE manager needs to recruit and retain the best technical professionals that can be found. These should have at least an associate's degree in electronics or biomedical technology, or equivalent military training. It would be desirable to have a bachelor's degree and, for the managers, some business management training. When recruiting, it is highly recommended that the hospital give preference to those with certification as a *certified biomedical engineering technician* (CBET), *certified radiology equipment specialist* (CRES), *certified laboratory equipment specialist* (CLES) or *certified clinical engineer* (CCE) from the International Certification Commission (ICC)[14] or the Healthcare Technology Certification Commission (HTCC)[15], as these professionals have proven to their peers that they have the knowledge, experience, and skills to fulfill their responsibilities. Obviously, certification is only a basic requirement, additional education and continuous training are essential to keep CE professionals current with technology advancement.

Once hired, CE staff needs to receive appropriate orientation before starting to perform their duties. The orientation should include not only an introduction to hospital structure and managers but also all the policies and procedures that need to be observed. Safety training and compliance to applicable regulations on worker safety and health hazards should also be included. After starting to work, like other healthcare workers, CE staff is required by regulation and accreditation standards to be evaluated periodically for technical competency. It would be wise to extend this evaluation to administrative abilities and communication skills. Depending on the results of these assessments, additional training or orientation should be prescribed and implemented. Whenever possible, younger professionals should be mentored by more experienced ones so the former can learn how to improve themselves and grow their careers.

4.4 SERVICE PROVIDER MANAGEMENT

Wang et al. [2008] reported that the total expenditure on service contracts for medical equipment is often 50% or more of the hospital's total expenditure on medical equipment maintenance and man-

[14]Visit its website at `http://www.aami.org/certification/index.html`, for more information.
[15]Visit its website at `http://www.accefoundation.org/certification.asp`, for more information.

agement, without considering expenses related to outside services paid without signed agreements. So careful consideration on who should be providing the service is one of the most important duties of any CE Department. This process is sometimes called *service delivery strategy* planning.

In principle, there are three distinct options for servicing medical equipment using one of the following labor sources:

(i) onsite staff[16]

(ii) the manufacturer, and

(iii) an alternative vendor.

The best option can vary from one piece of equipment to another due to a combination of the following factors:

- Technical competency: who is competent to perform the services needed (i.e., has been trained)
- Technical documentation: whether manuals, information, and/or software is available
- Parts and tools: whether the necessary parts and tools are available
- Response time: who can respond quicker and, most importantly, resolve the problem
- Cost: who provides the lowest cost
- Quality: who provides consistent and reliable quality

Although, in theory, the first option is best in most cases because it can be the first to respond and the cost is already budgeted, often factors such as lack of documentation, training, or inability to access parts or software prevent the onsite staff from performing the service needed. The representatives of the manufacturers and alternative vendors have to travel from their current locations to the hospital, thus slowing the response. Furthermore, as these entities have to keep field staff available even though the demand is unpredictable, their costs are inherently higher than the first option. Therefore, it is highly recommended that the CE Department performs a careful analysis of the service alternatives by determining which of the three options are available and, then, which one offers the best value in terms of quickest solution at the lowest possible cost. One alternative that should not be ignored is a combination of option (i) with either (ii) or (iii), i.e., the onsite staff responds first to all calls and requests assistance of the other when the former exhausted its abilities to solve the problem. This combination tends to result in prompt response (which often is the most important aspect for clinical users) while reducing unnecessary travels from the latter. Furthermore, experience shows that the majority of service calls can be resolved easily because they are related to user challenges and failures of disposable products, accessories, or environmental issues.

[16]Included in this category are hospital employees, hospital employees supervised by a contractor, and contractor staff.

When it is not possible to perform equipment service with onsite staff, CE managers need to verify the qualification and cost effectiveness of external service providers. Although the qualification of the manufacturer seldom needs to be verified (because it is under FDA scrutiny), the quality of its field service staff needs to be monitored. Problems with individual service representatives should be reported to the manufacturer management hierarchy as quickly as possible, with copies to the manager of the respective clinical department. Such monitoring is now required by some accreditation organizations as part of vendor review.

On the other hand, the qualification of third party, alternative providers of service and parts is quite uneven. Some can be very good and cost effective, while others are "cheap" at best. For this reason, it is essential that CE managers carefully evaluate every alternative provider of service and parts, not only initially but also periodically. One good practice is to pre-qualify alternative vendors by asking for information on the following topics:

1. Business license and registration
2. Proof of adequate general liability and vehicle insurance coverage
3. Implementation of a quality system (e.g., ISO 9001: 2008, FDA Quality System, CE Mark, etc.)
4. Evidence of staff qualification and competency verification
5. In the case of parts or accessories, samples for testing or test reports produced by reputable, independent agencies
6. Business and technical references

After reviewing the information above, the CE manager still needs to spot check the material and services provided. Furthermore, at least once per year the overall vendor performance should be evaluated and the decision to continue the business relationship reviewed.

4.5 FINANCIAL MANAGEMENT

One of the biggest challenges for the CE manager is to have detailed control over the finances. The CE manager should not only control the budget for his/her department but have oversight of all expenses related to the maintenance and management of all medical equipment. Too often, the CE Department budget only covers the labor expenses of its staff, supplies and basic replacement parts, and certain miscellaneous expenses. The rest, such as expensive replacement parts (e.g., CT tubes and other glassware), service contracts, and "extended warranties," are often recorded in the budgets of the clinical departments that "own" the equipment. This not only distorts any possible benchmarking efforts [Wang et al., 2008] but also reduces the ability of the CE Department to find the most cost-effective manner to support medical equipment.

A detailed discussion of how the CE expenses should be managed is beyond the scope of this chapter and the reader is referred to a chapter dedicated to this subject in another book [Wang, 2004]. Suffice to say here that the financial performance of the CE Department should be one of the basic elements of performance assessment (see Section 5.2.1 below) and needs to be monitored carefully by the hospital executive who is responsible for overseeing this department.

4.6 RELATIONSHIP WITH OTHER DEPARTMENTS

As mentioned above, CE staff needs to be technically competent. However, it is just as important that they have the proper customer-service attitude and skills. This is because the healthcare industry is unique in the sense that it depends on highly sophisticated equipment to delivery safe and effective care, yet most of the equipment users are not very technically savvy, if not downright scared of technology. This contradiction is rooted in the cultural tradition that most of the youngsters who do not like mathematics and physics tend to choose a career in life sciences and *vice versa*. On the other hand, CE staff typically is composed of quiet and introverted persons who often do not communicate well with non-technically oriented people. Thus, one of the main challenges in staff management is to train CE staff on customer service and communication skills, while continuing to encourage them to grow their technical knowledge and expertise.

To overcome the communication barrier and not let individual CE staff find his/her own way to address this issue, the CE manager—together with other hospital leaders—needs to institute certain routines that foster more frequent and fluid communication between CE and clinical staff. Successful approaches adopted by CE professionals in various organizations include the following:

1. Periodic "rounds" with key clinical supervisors and managers
2. Automated email updates on status of equipment in repair
3. Advance notice on equipment scheduled for inspection or preventive maintenance
4. Open-house, sponsored coffee breaks, and other social activities to promote mutual awareness and improved communication
5. Clinical user feedback in various forms, both unsolicited and periodic surveys
6. Participation in committees, task forces, volunteering events, etc.
7. Periodic (monthly, quarterly, etc.) reports to administration and clinical departments regarding accomplishments and challenges

4.7 INTERACTION WITH INFORMATION TECHNOLOGY

The accelerated introduction of electronic health records (EHR) in hospitals encouraged by CMS requires the transfer of data from many kinds of medical equipment into EHR. This demands close interaction between the Information Technology (IT) and CE departments.

Besides facilitating the transfer of data between medical equipment and the EHR, the CE Department needs to cooperate with IT to ensure compliance with HIPAA Security and Privacy Rules as described above. Furthermore, CE staff needs to verify that medical equipment connected to the IT network is properly protected against virus and malware, especially if the network is connected to the Internet.

A good guidance for the proper interaction between IT and CE is provided by the ANSI/AAMI/IEC 80001-1 standard (ANSI/AAMI/IEC, 2010b). Part 1 of this standard provides the roles, responsibilities, and activities of the stakeholders involved in the risk management of IT networks that incorporate medical equipment.

4.8 PERFORMANCE MONITORING AND CONTINUAL IMPROVEMENT

Monitoring and improvement of maintenance services were discussed above. However, since maintenance is not the only (or even most important) attribution of the CE Department, a global monitoring of the department's performance is needed to find opportunities for continual improvement.

One way to monitor the overall performance of the CE Department is to use the balanced scorecard approach proposed by Kaplan and Norton [1996] and widely adopted in a variety of industries. Figure 4.3 shows an example of a balanced scorecard for a CE Department. The four dimensions suggested are: (i) operational performance, (ii) staff learning and growth, (iii) user perspective, and (iv) financial performance [Wang et al., 2008]. Within each dimension, performance indicators need to be selected and monitored. Deficiencies detected by the indicators should be analyzed carefully to determine their root causes, so appropriate preventive and corrective actions can be taken to correct the problem and prevent future reoccurrence.

In order to achieve continual improvement, it is not enough to correct and prevent deficiencies. It is necessary to uncover issues that are not being monitored or measured. This requires periodic discussions (rounding) with all the stakeholders, i.e., clinical user departments, other support departments, administration, finance, etc., as well as learning from other hospitals and professional colleagues (see benchmarking discussion below). These inquiries will help to determine new performance indicators that need to be added (and sometimes the retiring of unnecessary ones), so the CE Department performance can be continually improved.

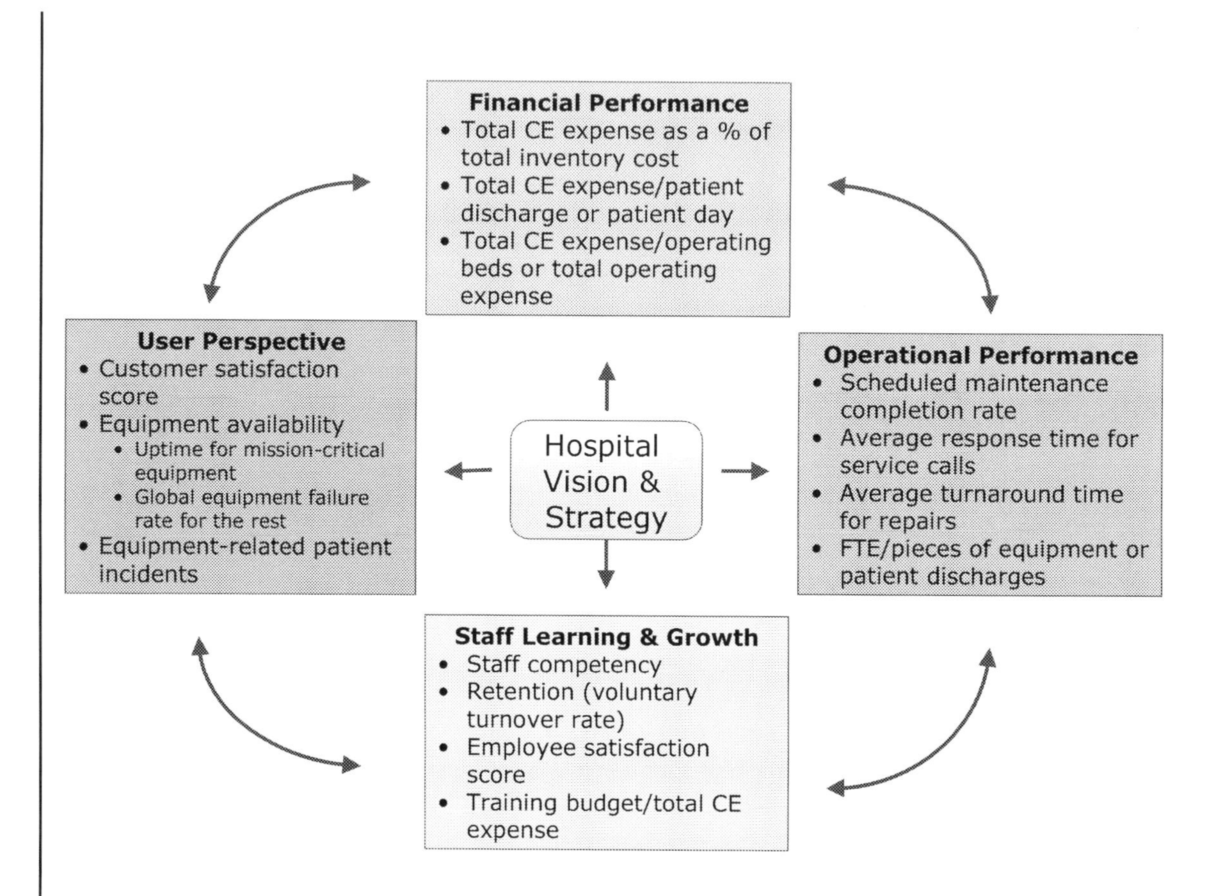

Figure 4.3: A sample balanced scorecard for a CE Department. The scorecard is divided into four dimensions. In the operational performance dimension, process indicators such as scheduled maintenance completion rates, average service call response time, etc., are used to gauge the implementation of defined service strategies. In the financial performance dimension, indicators such as total CE expense as a percentage of total inventory acquisition cost and total CE expense per patient discharge are used to monitor service efficiency. In the user perspective dimension, various indicators such as customer satisfaction score, uptime for mission critical equipment, and global failure rate for other equipment are used to assess service effectiveness. Finally, in the staff learning and growth dimension, indicators such as staff retention (turnover) rate and employee satisfaction score are used to measure staff motivation and growth potential. By itself, none of the four dimensions is sufficient to provide an accurate picture of a CE Department. Together, they provide a comprehensive assessment of the department and, thus, provide insight that can assist the CE manager to make corrections and improvements to foster growth and evolution.

CHAPTER 5

Performance Management

One of the biggest challenges of CE managers and senior hospital leaders who manage or oversee CE departments but do not have prior CE experience is how to measure its performance and compare it to other hospitals or service providers. Many of the performance indicators discussed above are too detailed and, most importantly, do not have references for comparative assessment. This section provides some fairly easy yet effective means to measure and compare performance.

5.1 MAINTENANCE EFFECTIVENESS

As mentioned above (Section 3.4.3), completion rate of scheduled maintenance by itself is not a good measure of maintenance effectiveness just as class attendance is not a reliable measure of student learning. While the failure cause coding process shown in Table 3.1 is a good way to measure and compare the effectiveness of maintenance processes (both scheduled and unscheduled), its implementation requires training and discipline of the CE staff, as well as analytical skills of CE leadership. This combination is likely not available everywhere but in the best-organized CE departments. For this reason, a simplified way to measure and monitor maintenance effectiveness is needed.

The ultimate goal of maintenance is reliability and safety, i.e., equipment should not fail often and when it does fail, it should be promptly repaired and returned to service; furthermore, it should always be safe for both patients and users. Therefore, a way to measure maintenance effectiveness is to determine how often equipment fail (i.e., failure rate) and how quickly it is repaired (i.e., turnaround time). Instead of using two indicators, an indicator that would be the ideal measure for reliability is "uptime," i.e., the fraction of time the equipment is available for use versus the total amount of time it is planned to be available. For example, a CT scanner in the Diagnostic Imaging Department may be planned for 10 hours of use per weekday, so its planned use time in a year is 52 weeks * 5 days/week * 10 hours = 2,600 hours. If the amount of actually available hours for use, after deduction for downtime due to failures and repairs, were 2,000 hours, then the uptime would be 76.9%.

Unfortunately, uptime is not easy to measure for the vast majority of equipment because it would require diligent logging of the dates and times when the equipment failed and when it was repaired and returned to service. Unless it is a mission-critical piece of equipment or one that has few, if any, backup, users frequently just put a note on the equipment stating "broken" and leave it aside, instead of calling CE promptly for repair. CE staff typically open and close a workorder after repairing the equipment, so the total elapsed time of the workorder is not an accurate measure of the downtime or turnaround time [Wang et al., 2012]. This explains why downtime is only adopted for major pieces of equipment (e.g., MRI, linear accelerator, CT, automated laboratory chemistry analyzer, etc.) for which time logging is part of the user routine due to high demand.

For the rest of medical equipment, the more practical approach is to use failure rate as the primary reliability measure[17]. While failure rate can be measured for individual pieces of equipment, it is not practical to deal with it when there are over 5-10,000 pieces of equipment in a hospital[18]. On the other extreme, a single failure rate for all medical equipment (termed "global failure rate" by Wang et al. [2006b]) could be too ambiguous for any decision. Therefore, it is recommended to start measuring failure rates for the main categories of equipment (i.e., biomedical, imaging, oncology, and clinical laboratory) and then drill down from there to specific groups by function (e.g., diagnostic, monitoring, therapeutic, etc.), purpose (e.g., resuscitation, physical therapy, surgery, intensive care, etc.), and finally down to the brand and model level. At higher levels, accuracy is sacrificed due to the large variability that is included in the group but less time is needed to acquire enough measurements for statistical analyses. As one drills down, the accuracy increases but more time is needed for data collection. As a rule of thumb, it is best to start at a fairly high level and dive into smaller groups when there is a need or desire to understand better the underlying causes and opportunities for improvement.

Table 5.1 shows results of analyses of data collected from hundreds of hospitals that can be used as a reference for evaluating individual CE departments' maintenance effectiveness [Wang et al., 2006b, 2008, 2011].

5.2 MAINTENANCE EFFICIENCY

While the maintenance effectiveness discussed above provides a way to measure the outcome of the CE Department, it does not measure how efficiently the resources are used to achieve the desired outcomes. Like any other enterprise, the CE Department has to be both effective and efficient in order to be of value for the healthcare organization it supports or, to put it more bluntly, to survive and thrive.

There are two basic ways to measure CE efficiency. The first one is to use the overall cost of maintaining all the medical equipment within the healthcare organization as a way to determine how efficiently the financial resources are being used to keep the inventory safe and reliable. The second is to use the amount of full-time equivalent (FTE) staff deployed by the CE Department as a way to determine labor productivity. These two measures are discussed below.

[17] One precaution that is needed is to isolate failures that are truly due to equipment and not those caused by use errors (also known as "cannot duplicate"), peripherals, and network issues (see Section 3.4.3 above).

[18] On the other hand, individual failure rates can be useful for determine the need for equipment replacement, as equipment that fails often may be reaching its useful life or there are other failure causes that need attention.

Table 5.1: Ranges of annual failure rate for the main medical equipment categories and some commonly found equipment groups collected from acute care hospitals

	INDICATOR	ANNUAL FAILURE RATE RANGE
CATEGORY	All biomedical equipment	0.6–0.8
	All imaging equipment	5.8–7.8
	All laboratory equipment	0.6–0.8
GROUP	Physiological monitoring system	0.35–0.55
	Ventilator, adult, pediatric, neonatal & portable	0.50–0.75
	Blood warmer	0.07–0.25
	Anesthesia machine	0.50–0.70
	Infusion pump, single and multiple channel	0.40–0.85
	Syringe infusion pump	0.17–0.37
	Patient-controlled analgesia pump	0.20–0.35
	Vital signs monitor	0.55–0.75
	Ultrasound scanner	0.35–0.55
	Electro-surgical unit	0.13–0.33
	Defibrillator	0.29–0.49
	Infant warmer	0.20–0.40
	NIBP monitor	0.60–0.85
	Infant scale	0.15–0.35
	Enteral feeding pump	0.16–0.36
	Pulse oximeter	0.20–0.40
	Blanket warmer	0.17–0.37
	Patient scale, floor model	0.16–0.36
	Sequential & intermittent compression device	0.23–0.43

5.2.1 FINANCIAL EFFICIENCY

Considerable amount of discussions have been held over the last three decades on how to measure financial efficiency for CE departments (see, e.g., David and Rohe [1986]; Furst [1986, 1987]; Bauld [1987]; Johnston [1987]; Frize [1990a,b]). Until 2008, the most promising metric was the "cost of service ratio"—CSR, i.e., the total equipment maintenance cost as a percentage of the total acquisition cost of medical equipment, proposed by Cohen [Cohen et al., 1995; Cohen, 1997, 1998]. In 2008, Wang et al. [2008] published an analysis of data collected from 253 acute care hospitals and concluded that CSR is not the only or best metric. Actually, several other denominators have better statistical correlation with the total maintenance cost and more accurate values, such as:

1. Total operating expenses of the healthcare organization
2. Total and adjusted patient discharges
3. Total and adjusted patient days
4. Total amount of operating ("staffed") beds

This conclusion is not totally surprising because very few hospitals have carefully tracked the acquisition cost of all its equipment purchases. Furthermore, some included the cost of accessories, shipping, and taxes, while others did not. Many hospitals do not include equipment with unit cost less than $5,000 in their capital asset while others use a much lower threshold. In contrast, the four parameters listed above are carefully scrutinized because they are audited by CMS and state agencies when the hospital requests reimbursements from Medicare and Medicaid funds, as well as by insurance companies.

A financial model was created using the four parameters listed above and data collected from 154 acute care hospitals that contracted ARAMARK Healthcare Technologies to maintain and manage its medical equipment. Figure 5.1 shows a comparison of actual total CE expenses collected from 275 hospitals through Thomson Reuters (now Truven Health Analytics[19]) ACTION O-I® program with ranges computed from the financial model. This comparison shows that the majority of ACTION O-I® members have total CE expenses within the range predicted by the model. The outliers are most likely caused by erroneous accounting of CE expenses[20], although the major teaching hospitals could have higher expenses due to their tendency to acquire the most sophisticated equipment and in larger quantities.

5.2.2 STAFF PRODUCTIVITY

Productivity has been equally debated for several decades without consensus on how it should be measured. Until recently, many CE professionals believed that a suitable measure would be the ratio of divided by the number of paid hours [ASHE, 1982; Bauld, 1987; Fennigkoh, 2004; Kawohl et al., 2005]. Unfortunately, the number of hours documented on workorders is not audited by anyone and, thus, subject to manipulation by CE staff [Wang et al., 2012].

[19] See `http://www.truvenhealth.com/` for more information.

[20] One of the accounting challenges that some hospitals face is the need to "normalize" all the expenses associated with medical equipment maintenance and management from various departments to the CE Department. In many hospitals, the CE Department budget only covers its staff wages and benefits, miscellaneous supplies and consumables, and overhead; whereas the cost of repair parts (some of which are rather costly like the X-ray tubes) are charged back to the respective clinical user departments. Without proper "normalization," total CE expenses would be substantially lower than the actual amount spent on equipment maintenance and management (see Wang et al. [2012], for more details and actual data).

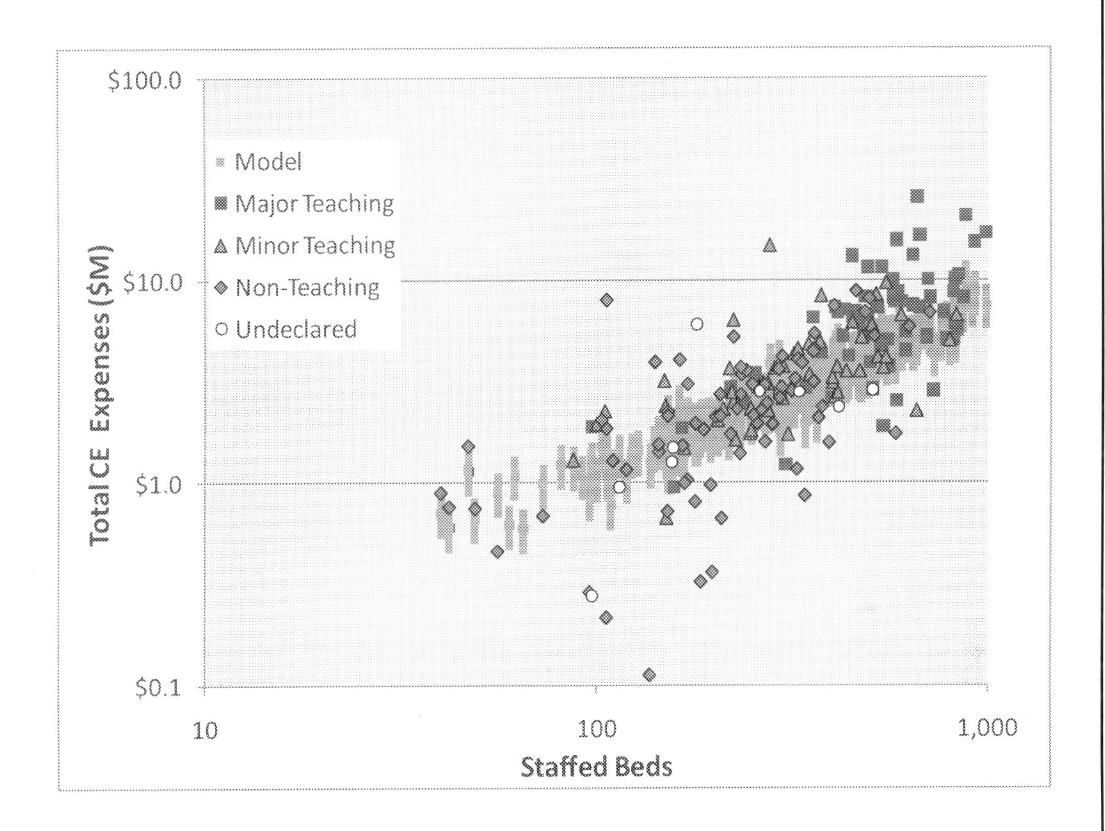

Figure 5.1: Comparison of actual total CE expenses collected by Thomson Reuters through its ACTION O-I® program from 275 hospitals in 2010 (with symbols showing hospitals with different teaching characteristics) and ranges (vertical gray bars) computed using a multi-dimensional model based on actual data collected by ARAMARK Healthcare Technologies in the same period from 154 acute care hospitals that retained this company to provide CE services. Each vertical bar is centered on the CE expense calculated by the model, while the size of the bar indicates the range of the values estimated. Thick gray bars were intentionally used for the estimated ranges so the actual CE expense data points would not be obscured.

The Bureau of Labor Statistics (BLS) defines labor productivity as the ratio of the output of goods and services to the labor hours devoted to the production of that output [BLS, 2012]. Since it is difficult to attribute a monetary value to the service provided by the CE Department and the total number of hours documented on workorders is unreliable, Wang et al. [2012] examined the following alternative parameters:

1. Total number of operating beds
2. Total (adjusted) patient discharges
3. Total (adjusted) patient days
4. Total hospital operating expense
5. Total number of pieces of medical equipment
6. Total acquisition cost of medical equipment
7. Total number of (scheduled and unscheduled) maintenance workorders

These parameters were used to build a staffing model similar to the financial model above using data collected from 154 acute care hospitals managed by ARAMARK Healthcare Technologies. Figure 5.2 shows a comparison of actual full-time equivalent (FTE) collected from 275 hospitals through Thomson Reuters (now Truven Health Analytics) ACTION O-I® program with ranges computed from the staffing model. This comparison shows that the majority of ACTION O-I® members have total FTEs within the range predicted by the model.

It needs to be stressed that unlike financial efficiency, labor productivity is not a good way to judge the overall cost effectiveness of any CE Department by itself alone. First, labor expense is typically around 15-20% of the total equipment maintenance and management cost, assuming proper "normalization" has been performed [Wang et al., 2008, 2012]. So variations in FTE are often not easily detectable. Furthermore, the size of the staff depends strongly on the management model adopted by the hospital. On one extreme, it is possible to have only one person managing the CE Department by placing all equipment on service contracts. This approach has very high apparent productivity but likely spends more than others will; furthermore, the service response time is probably unsatisfactory, unless there is ample local availability of contract service representatives. On the other extreme, a very large CE team could be assembled to maintain and repair everything themselves down to the level of components. This approach may spend less than others but the quality and turnaround time are likely unacceptable if not potentially unsafe for patients and users.

5.3 PERFORMANCE BENCHMARKING

The prior discussions on maintenance effectiveness and efficiency have hinted that performance benchmarking with similar organizations is something that must be considered for those who are

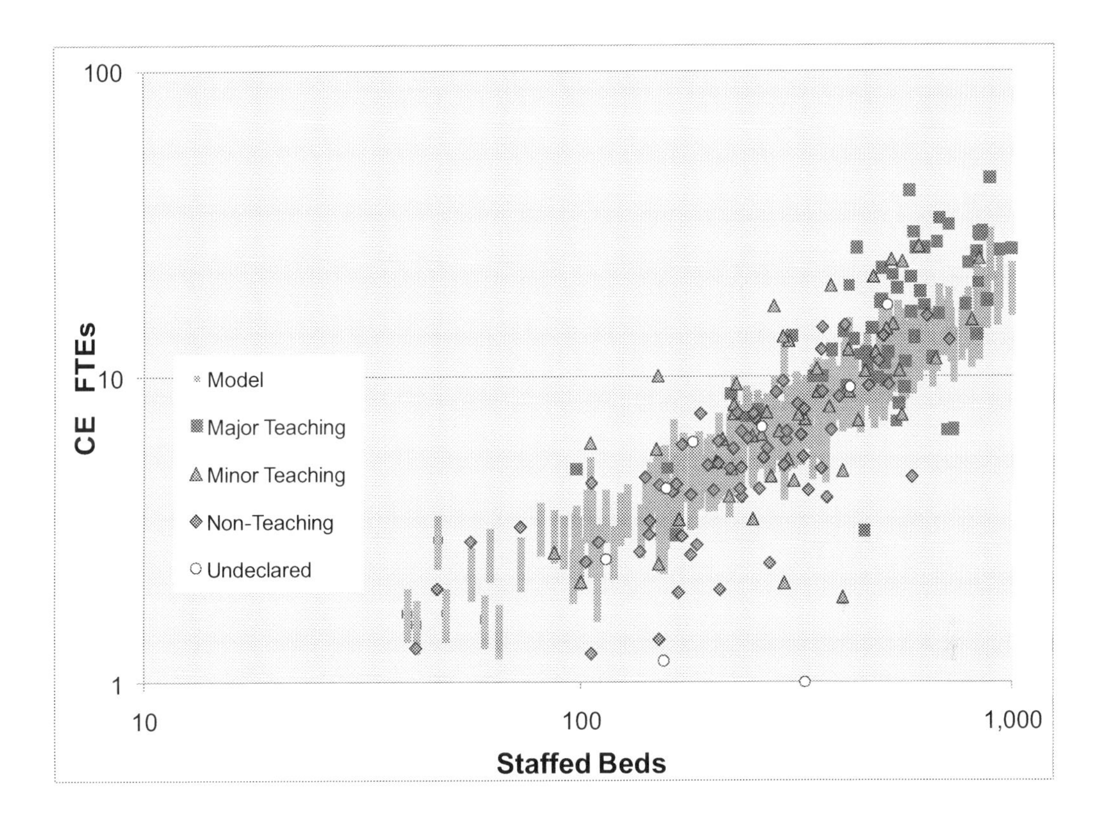

Figure 5.2: Comparison of actual FTE values collected by Thomson Reuters through its ACTION O-I® program from 275 hospitals in 2010 (with symbols showing hospitals with different teaching characteristics) and ranges (vertical gray bars) computed using a multi-dimensional model based on actual data collected by ARAMARK Healthcare Technologies in the same period from 154 acute care hospitals that retained this company to provide CE services. Each vertical bar is centered on the FTE value calculated by the model, while the size of the bar indicates the range of the values estimated. Thick gray bars were intentionally used for the estimated ranges so the actual FTE data points would not be obscured.

responsible for the management or oversight of CE departments. Although CE benchmarking has been discussed for well over two decades and sometimes the debates have been quite heated, very few valid benchmarks have been found and accepted by CE professionals [Furst, 1987; Cohen et al., 1995; Hutchins, 2006; Maddock, 2006; Gaev, 2010]. As pointed out by Wang et al. [2008], the data available only allow ballpark comparisons instead of accurate benchmarking.

Currently, there are three organizations providing CE benchmarking services. They are listed below in alphabetic order with respective Internet website links:

1. Association for the Advancement of Medical Instrumentation—AAMI: AAMI's Benchmarking Solution (ABS), `http://www.aami.org/abs/index.html`;

2. ECRI Institute: BiomedicalBenchmark^{TM},
   `https://www.ecri.org/Products/Pages/BiomedicalBenchmark.aspx`;

3. Truven Health Analytics (formerly Thomson Reuters Healthcare): Action O-I® Operational Performance Improvement Solution,
   `http://thomsonreuters.com/products_services/healthcare/healthcare_products/a-z/action_oi/`

The first two are dedicated to CE benchmarking while the last one covers the entire hospital. On the other hand, Truven claims to have >750 contributors, while the other two have less than 200. If the hospital is already a subscriber to Truven, then it should consider using Action O-I®; otherwise, the first two could be considered as alternatives.

Unfortunately, none of these benchmarking solutions provides the comprehensive evaluation outlined in Figure 4.3. The Financial Performance quadrant is well addressed by maintenance efficiency and most of Operational Performance elements are already being monitored. On the other hand, while maintenance effectiveness is an objective measure of maintenance outcome, it is only one of the elements of User Perspective. If clinical users' perception of the CE Department is poor for whatever reason, it will be difficult for it to grow and gain attention and support of senior executives. The most important and often overlooked quadrant is Learning and Growth. It is the CE manager's top priority to recruit, retain, motivate, and enhance the qualification of CE staff.

CHAPTER 6

Discussion and Conclusions

Readers may be surprised by the amount of attention lavished on regulatory framework. Although most CE managers and hospital senior leaders with oversight responsibility probably know most of the applicable laws, regulations, and standards, it is believed that the best way to be able to supervise properly a CE Department is to know why it is doing what it is doing instead of simply doing what has been done previously. As explained eloquently by Sinek [2009], workers not only perform better but are inspired to do their best if they are provided with the reason these tasks need to be performed by their leaders.

The material provided in this book applies to CE departments within the majority of acute care hospitals. However, since no two hospitals are identical, adjustments have to be made for individual cases. While most of the necessary adjustments are likely to be intuitive for seasoned administrators, some may deserve a bit more attention and, thus, are discussed below.

6.1 CHALLENGES RELATED TO SIZE, ACUITY, AND GEOGRAPHICAL LOCATION

While basic epidemiological characteristics are not strongly dependent on geographical location and population density, hospitals do vary significantly from each other due to a combination of factors, such as size, patient acuity, profit orientation, teaching characteristics, geographical location, etc. These differences impact CE departments directly and indirectly.

Small, rural hospitals tend to be more limited in resources and, in addition, have difficulties in recruiting top talents from metropolitan areas. These factors typically prevent these hospitals from establishing first-rate CE departments. Hospitals located in remote areas have the additional disadvantage of lack of support by manufacturers and third-party service vendors. This translates to slow response time and even longer repair turnaround time. For these reasons, executives of these hospitals often seek assistance from independent service organizations (ISOs) that can marshal nationwide resources to support a small CE Department.

On the other extreme, very large, major teaching hospitals tend to have a large quantity of sophisticated equipment, as well as customization needs to suit research interests. Instead of focusing on maintenance efficiency, these hospitals often spend more than necessary because of the challenges in controlling equipment incorporation and utilization well. Equipment is sometimes purchased simply to satisfy a prestigious physician or surgeon who may only use the equipment occasionally. Since many of the prominent hospitals receive generous donations and endowments, they often are not very concerned about cost effectiveness.

For-profit hospitals tend to be the most demanding of their CE departments. In order to be profitable, these hospitals have high equipment utilization and only replace equipment when there is a clear necessity, be it a strategic competitive advantage or when the existing equipment fails too often or is too expensive to repair. The high utilization translates into low availability of equipment for service and demands for quick turnaround of repairs. The intense usage and slow replacement result in a high failure rate and, thus, frequent repairs. Thus, it is essential that these CE departments are well staffed with highly skilled and experienced service personnel.

The challenges above help explain some of the wide differences found in financial and productivity analyses and, thus, should be considered by CE managers and their overseers in evaluating the performance of CE departments.

6.2 CHALLENGES RELATED TO CULTURAL TENDENCIES

As mentioned above (Section 4.6), CE staff is typically composed of persons who are reserved, introverted, and passionate about their jobs. They feel more comfortable around machines than people and strongly believe they are champions of patient safety. Sometimes it is not surprising that they are seen as "stubborn" or "zealots" for defending their beliefs and willingness to sacrifice their personal gains for the benefit of the patients and their clinical colleagues. It is not uncommon to find some CE professionals who are adverse to commercial organizations and salespersons because of the profit orientation of the latter.

In addition to often inadequate communication skills, CE professionals have a tendency of losing sight of the big picture when they focus on individual issues. It is not surprising to find CE technicians spending several hours to save a few dollars when searching for repair parts, while the flow of patient care is halted. Similarly, some will waste inordinate amounts of resources to ensure 100% safety for everything under their responsibility, forgetting that perfection is a reasonable goal but not a realistic result. These shortcomings can be addressed by training, counseling, and mentoring. Those who can overcome these challenges are typically promoted to managers, directors, or even higher ranks.

Overseers of CE departments who understand these cultural tendencies know that they are dealing with people who are honest, dedicated, and willing to understand objective and rational discussions. Once the manager or overseer earns the trust of a CE professional, he or she can be confident that the CE professional will perform beyond his/her duty regardless how challenging the situation may become.

6.3 EVIDENCE-BASED MAINTENANCE

As mentioned above in Section 5.1, the ultimate goal for medical equipment maintenance and management is to enhance reliability and safety for both patients and users. Other healthcare professionals probably think this goal seems quite similar to that of healthcare or medicine, i.e., to keep patients healthy and safe. Actually, if one substitutes equipment for patient, almost all the principles and methods of healthcare and medicine are applicable to CE.

In particular, one healthcare methodology that is especially relevant to CE is evidence-based medicine—EBM [EBMWG, 1992]. Medical practitioners need to supplement their medical education and clinical training with the results of the most recent comparative effectiveness studies, such as those conducted using randomized clinical trials on drugs, devices, and procedures. Likewise, CE professionals need to supplement their engineering education and technical training with up-to-date results of the maintenance effectiveness studies, which evaluate critically different maintenance strategies, procedures, and frequencies. This methodology has been termed evidence-based maintenance—EBMaint [Wang, 2007, 2008; Wang et al., 2010a,b, 2011], which can be defined as a continual improvement process that analyzes the effectiveness of maintenance resources deployed in comparison to outcomes achieved previously or elsewhere and makes necessary adjustments to maintenance planning and implementation.

In other words, instead of simply following laws, regulations, codes, standards, industry practices, and manufacturers' recommendations, CE managers need to learn from their colleagues the results of the latest comparative effectiveness studies. For example, Wang et al. [2010b] showed that the reliability—as measured by failure probability distributions—of certain types of medical equipment (e.g., pulse oximeters, vital signs monitors, and radiant infant warmers) is virtually independent of the maintenance strategy adopted. These results suggest that periodic inspection of certain types of equipment does not enhance reliability or safety and, thus, is a waste of resources. Another example is the comparison of maintenance procedures recommended by the manufacturer versus those developed by the CE teams themselves. The former are typically written by design engineers who have limited service experience themselves and tend to reproduce production-line test procedures that are excessively detailed and time consuming; whereas the latter are a condensed version of manufacturers' recommendations generated after performing numerous cycles of SM and, thus, much shorter and to the point. A preliminary comparison of reliability outcomes show no detectable differences for several types of medical equipment, such as pulse oximeters, vital signs monitors, telemetry transmitters, aspirators, respiratory humidifiers, enteral feeding pumps, patient scales, and multi-parameter patient monitors [Wang et al., manuscript in preparation].

Even more important than effectiveness comparison of maintenance strategies, procedures, and frequencies is the insight provided by the EBMaint studies. Analyses of failure probabilities of 22 types of equipment showed that only ~3% of the failure causes could be affected directly by CE maintenance activities, while ~38% could be indirectly affected by CE staff if they were involved with the planning of new equipment acquisitions or replacement of existing equipment, especially in the crucial step of evaluating human factors together with the clinical users before selecting equipment to be acquired, as well as with the technical guidance and training for the clinical staff on proper use and prompt detection of erroneous uses of equipment [Wang et al., 2011]. This conclusion is consistent with the patient incident ("sentinel event") statistics accumulated and analyzed by TJC [2012a]. While medical equipment related incidents are the tenth most common type of incidents, root-cause analyses of those incidents show >82% of them are not brought about by equipment failures but by other causes such as human factors, leadership, communication, assessment, etc. [TJC, 2012a]. It is even unclear whether any significant portion of the causes is truly related to equipment failures [Wang et al., in press]. A good example is the infusion-pump related incidents. Studies have shown that the vast majority of them are caused by human factors and other clinical issues [AAMI, 2010a; Wang et al., 2011]. Therefore, it does not make sense to try to solve this issue by asking CE staff to inspect pumps more frequently. In other words, as the goal of equipment maintenance and management is to enhance reliability and safety for patients and users, the most cost effective way to deploy CE resources is not in equipment maintenance but in technology management.

EBMaint is gradually gaining acceptance among CE professionals and the attention of regulators. For example, in the December 2011 revision of its equipment maintenance guidelines, CMS [2011] stated, "if the hospital is adjusting maintenance activity frequencies below those that are recommended by the manufacturer, such adjustments must be based upon a systematic evidence-based assessment." While other parts of this new set of guidelines have been hotly debated, several health systems and a well-known patient-safety organization have endorsed EBMaint as the preferred method to determine maintenance processes [Douglas, 2012]. Like evidence-based medicine, it will take time for CE professionals to change their current practice and adapt to the new reality, but eventually EBMaint will become the standard method of managing equipment maintenance.

6.4 CONCLUSIONS

In spite of its small size and budget compared to the rest of the clinical and non-clinical departments, CE departments are vital resources for all healthcare organizations. Without safe and reliable medical equipment, it is impossible to deliver care. Furthermore, if the CE Department is not properly managed and does not comply with applicable laws, regulations, and standards, the hospital runs the risk of losing the significant portion of its revenue that is derived from Medicare and Medicaid reimbursements (if not from other insurance companies). On the other hand, a well managed CE Department not only can help the hospital to earn more revenue (by making equipment more reliable and, thus, available for patient care) but also can reduce the amount of revenue the hospital needs to generate in order to maintain the equipment in safe and good operating order[21].

While CE staff is required to be highly competent in technical areas, CE managers and senior leaders who oversee these departments are not required to have in-depth technical competency or experience. However, the latter need to understand and appreciate the reasons why certain equipment maintenance and management tasks need to be performed, how to monitor and measure the performance of the CE team, and how its performance compares with other teams in similar organizations, so improvements can be made continually. It is hoped that the material provided in this book is helpful in providing these tools to CE managers and overseers.

[21] As the CE budget is roughly 1% of the hospital total expense [Wang et al., 2012], each dollar that the CE Department manages to save translates to $100 less the hospital needs to earn.

APPENDIX A

Sample Policies for a Clinical Engineering Department

The policies shown below only provide the principles that govern the Clinical Engineering (CE) Department in a hospital or healthcare system. Details of the processes used for implementing each policy are described in one or more procedures that may vary somewhat from one part of a system to another to accommodate different environments, patient demographics, or unique circumstances.

1. **Equipment Planning**

   CE Department shall assist the hospital with the planning and incorporation of new medical equipment and replacement of existing equipment.

2. **Equipment Acquisition**

   CE Department shall assist the hospital with the procurement of new and replacement medical equipment.

3. **Equipment Installation and Acceptance**

   The safety and performance of medical equipment shall be verified when it is received and installed (if applicable), regardless of ownership, before patient use. New equipment shall only be accepted if the requirements specified in the purchase order (e.g., operating and service documentation, user and service training) have been satisfied.

4. **Equipment Inventory**

   A current and accurate inventory of medical equipment shall be maintained by the CE Department.

5. **User Training**

   CE Department shall assist the hospital in determining training needs, sources, and methods for medical-equipment users, and with the coordination of technical utilization training.

6. **Equipment Maintenance**

   Maintenance of medical equipment, regardless of ownership, shall be planned and implemented using appropriate maintenance strategies to keep it safe and performing according to original functional specifications.

7. **Equipment Retirement and Disposition**

   CE Department shall assist the hospital with the retirement and disposition of unwanted medical equipment.

8. **Equipment Modification**

   Equipment safety and functional specifications shall not be altered by the CE Department or third parties under its supervision, unless directed by the respective manufacturer or the Food and Drug Administration (FDA) in a specific recall or upgrade action.

9. **Medical Device Recalls**

   Recalls issued by the manufacturer and/or FDA for equipment managed by the CE Department shall be monitored continuously and addressed promptly to reduce risks to patients and impacts on the delivery of health services.

10. **Medical Device Tracking**

    CE Department shall assist its customers, upon their request, in determining and implementing appropriate processes to track medical devices deemed "trackable" by the FDA.

11. **Patient Incident Investigation**

    CE Department shall assist the hospital in investigating patient incidents involving medical equipment and follow hospital procedures for reporting such incidents.

12. **Control of Documents**

    Documents shall be controlled in terms of creation, approval, revision, preservation, and disposal to ensure integrity and consistency. All medical equipment management policies and procedures shall be approved by the appropriate hospital committee or officer before implementation. The most current, approved version of each document shall be available to the staff, if relevant to his/her job responsibilities.

13. **Control of Records**

    Records shall be generated when work is performed to provide evidence of work completion. The records shall be controlled until they can be disposed per hospital record retention policy or procedure.

14. **Management of Test and Calibration Devices**

    Devices used for inspections, tests, measurements, and calibrations of medical equipment shall be inventoried, tracked, serviced, and calibrated (if applicable) at appropriate intervals.

15. **Vendor Management**

    Vendors, including manufacturers, and their services (parts and labor) shall be qualified, monitored, and supervised to ensure satisfactory quality and cost effectiveness.

16. **Financial Management**

The financial performance of the medical equipment management program shall be controlled.

17. **Customer Relationship**

The requirements and expectations of each user department shall be established at the start up of operations and reviewed periodically. Close and frequent communication with user department management shall be maintained by the CE Department management using mutually agreed upon communication methods. CE Department shall participate in and/or contribute to hospital committees and task forces related to medical equipment management and maintenance.

18. **Performance Monitoring and Continual Improvement**

The performance of the medical equipment management program shall be monitored and evaluated periodically using a combination of indicators and audits, with the purpose of finding opportunities for continual improvement. Continual improvement projects shall be implemented and their results evaluated.

19. **Staff Management and Development**

CE Department shall provide qualified and competent management and technical staff to perform management and maintenance of medical equipment. New employees and current CE Department employees reassigned to new positions shall receive initial job orientation covering both hospital and CE Department information, policies and procedures, and duties and responsibilities. Staff competency to perform duties and responsibilities shall be assessed periodically. Education, training, certification, and other opportunities to advance professional skills shall be offered to management and technical staff. The performance of management and technical staff shall be evaluated periodically.

APPENDIX B

Information and Data Sources on Internet

Listed below are information and data sources available (mostly free) through the Internet. This list is not exhaustive but contains the main government agencies and non-profit organizations that produce or provide regulations, standards, and guidance on health technology.

- United States Federal Agencies
  - U.S.-Agency for Healthcare Research and Quality (AHRQ): `http://www.ahrq.gov`
  - U.S.-Centers for Disease Control (CDC): `http://www.cdc.gov`
  - U.S.-Centers for Medicare and Medicaid Services (CMS): `http://www.cms.gov`
    * Conditions of Participation: `http://www.cms.gov/Regulations-and-Guidance/Legislation/CFCsAndCoPs/index.html`
    * Clinical Laboratory Improvement Amendments (CLIA): `http://www.cms.gov/Regulations-and-Guidance/Legislation/CLIA/index.html`
    * Clinical Laboratory Accreditation: `http://www.cms.gov/Regulations-and-Guidance/Legislation/CLIA/Accreditation_Organizations_and_Exempt_States.html`
  - U.S.-Code of Federal Regulations: `www.gpoaccess.gov/cfr/`
  - U.S.-Department of Health and Human Services (HHS), Office for Civil Rights
    * Health Information Privacy and Security (HIPAA): `http://www.hhs.gov/ocr/privacy/`
  - U.S.-Federal Emergency Management Agency: `http://www.fema.gov/`

- U.S.-Food and Drug Administration (FDA):
  http://www.fda.gov/default.htm
  * Biologics (blood and blood products):
    http://www.fda.gov/BiologicsBloodVaccines/default.htm
  * Clinical Laboratory Improvement Amendments (CLIA):
    http://www.fda.gov/MedicalDevices/DeviceRegulationandGuidance/IVDRegulatoryAssistance/ucm124105.htm
  * Medical devices:
    http://www.fda.gov/MedicalDevices/default.htm
  * Radiation-Emitting Products:
    http://www.fda.gov/Radiation-EmittingProducts/default.htm
- U.S.-National Institutes of Health (NIH):
  http://www.nih.gov
- U.S.-Occupational Safety and Health Administration (OSHA):
  http://www.osha.gov/
  * Hazardous materials:
    http://www.osha.gov/pls/oshaweb/owadisp.show_document?p_table=STANDARDS&p_id=10117
  * Bloodborne pathogens
    http://www.osha.gov/pls/oshaweb/owadisp.show_document?p_table=STANDARDS&p_id=10051
  * Ionizing radiation:
    http://www.osha.gov/pls/oshaweb/owadisp.show_document?p_table=STANDARDS&p_id=10098

• Accreditation, Standards, and Testing Organizations
  - American Association of Blood Banks (AABB):
    http://www.aabb.org/Pages/Homepage.aspx
  - American Osteopathic Association (AOA) HFAP accreditation:
    http://www.hfap.org/
  - American Society of Histocompatibility and Immunogenetics (ASHI):
    http://www.ashi-hla.org/
  - COLA (laboratory accreditation):
    http://www.cola.org/
  - College of American Pathologists (CAP):
    http://www.cap.org/apps/cap.portal

- The Joint Commission:
  http://www.jointcommission.org
- DNV Healthcare:
  http://dnvaccreditation.com/pr/dnv/default.aspx
- ECRI Institute:
  http://www.ecri.org
- MD Buyline:
  http://www.mdbuyline.com
- National Fire Prevention Agency:
  http://www.nfpa.org
- The Advisory Board Company:
  http://www.advisory.com
- Underwriters Laboratories:
  http://www.ul.com

• Professional and Trade Associations

- Advanced Medical Technology Association (AdvaMed formerly HIMA):
  http://www.advamed.org
- American College of Clinical Engineering (ACCE):
  http://accenet.org
- American Hospital Association:
  http://www.aha.org
- American Institute of Architects:
  http://www.aia.org/
- American Institute for Medical and Biomedical Engineering:
  http://www.aimbe.org
- American Medical Association:
  http://www.ama-assn.org
- American Society for Healthcare Engineering:
  http://www.ashe.org
- American Society for Healthcare Risk Management:
  http://www.ashrm.org
- American Society for Testing and Materials:
  http://www.astm.org
- Association for the Advancement of Medical Instrumentation:
  http://www.aami.org

  - Biomaterials Network:
    http://www.biomat.net
  - Engineering in Medicine and Biology Society of the IEEE:
    http://www.embs.org/
  - Healthcare Information and Management Systems Society (HIMSS):
    http://www.himss.org
  - Healthcare Technology Assessment International:
    http://www.htai.org
  - International Society of Biomechanics:
    http://isbweb.org/
  - Institute for Healthcare Improvement (IHI):
    http://www.ihi.org
  - Institute of Electrical and Electronics Engineers (IEEE):
    http://www.ieee.org
  - International Federation for Medical and Biological Engineering:
    http://www.ifmbe.org
  - Medical Equipment and Technology Association (META):
    http://www.mymeta.org
  - National Patient Safety Foundation (NPSF):
    http://www.npsf.org
  - National Electrical Manufacturers Association (NEMA):
    http://www.nema.org
  - Radiological Society of North America (RSNA):
    http://www.rsna.org
  - Regulatory Affairs Professionals Society (RAPS):
    http://www.raps.org
  - VA National Center for Patient Safety (NCPS):
    http://www.patientsafety.gov

- International Organizations
  - International Organization for Standardization (ISO):
    http://www.iso.org/iso/home.html
  - World Health Organization (WHO):
    http://www.who.int/en/
  - Pan American Health Organization/Organización Panamericana de Salud (PAHO):
    http://new.paho.org/index.php

- Medical Information Sources
  - Medical Articles (Medscape):
    `http://www.medscape.com`
  - Medline/PubMed:
    `http://www.nlm.nih.gov/bsd/pmresources.html`
  - WebMD:
    `http://www.webmd.com/`
- Trade Magazines and Publications
  - 24x7 magazine:
    `http://24x7mag.com`
  - Health Technology Management:
    `http://www.healthcaretechnologymanagement.com`
  - Journal of Clinical Engineering:
    `http://journals.lww.com/jcejournal/pages/default.aspx`
  - TechNation:
    `http://1technation.com`
  - Biomedical Instrumentation and Technology (AAMI):
    `http://www.aami.org/publications/BIT/`
  - Medical and Biological Engineering and Computing:
    `http://www.springer.com/biomed/human+physiology/journal/11517`
- Benchmarking Organizations
  - Association for the Advancement of Medical Instrumentation—AAMI: AAMI's Benchmarking Solution (ABS),
    `http://www.aami.org/abs/index.html`
  - ECRI Institute: BiomedicalBenchmark^{TM},
    `https://www.ecri.org/Products/Pages/BiomedicalBenchmark.aspx`
  - Truven Health Analytics (formerly Thomson Reuters Healthcare): Action O-I® Operational Performance Improvement Solution,
    `http://thomsonreuters.com/products_services/healthcare/healthcare_products/a-z/action_oi/`

Bibliography

Association for the Advancement of Medical Instrumentation – AAMI, (1999) ANSI/AAMI *EQ56:1999 – Recommended practice for a medical equipment management program*, Arlington, VA. Cited on page(s) 15

Association for the Advancement of Medical Instrumentation – AAMI, (2010a) Infusing Patients Safely, *Priority issues from the AAMI/FDA infusion device summit,* Arlington, VA. Cited on page(s) 1, 54

Association for the Advancement of Medical Instrumentation – AAMI, (2010b) ANSI/AAMI/IEC 80001–1:2010 Application of risk management for IT Networks incorporating medical devices—Part 1: Roles, responsibilities and activities. Cited on page(s) 41

Association for the Advancement of Medical Instrumentation – AAMI, (2011a) *A Siren Call to Action – Priority issues from the medical device alarms summit*, Arlington, VA. Cited on page(s) 1, 22

Association for the Advancement of Medical Instrumentation – AAMI, (2011b) Forum Recommends Unified Name for Field, available at `http://www.aami.org/news/2011/050211.forum.name.html`, accessed on 7/22/2012 (AAMI, 2011b). Cited on page(s) 2

American College of Clinical Engineering – ACCE, (2012) Clinical Engineer (Defined), available at `http://www.accenet.org/default.asp?page=aboutandsection=definition`, accessed on 7/22/2012. Cited on page(s) 2

Ackoff, R.L. and Rovin, S. (2005) *Beating the System: Using creativity to outsmart bureaucracies*, Maintenance causes failures, pp. 48–49, Berrett-Koehler Publishers, San Francisco. Cited on page(s) 23

American Society for Hospital Engineering – ASHE, (1982) Part 2: Determining Productivity, in Medical Equipment Management in Hospitals, ASHE, Chicago IL, pp. 5–7. Cited on page(s) 24, 46

American Society for Hospital Engineering – ASHE, (1996) *Medical Equipment Management in Hospitals*, ASHE, Chicago IL. Cited on page(s) 24, 35

Bauld, T.J. (1987) Productivity: Standard terminology and definitions, *J Clin Eng* 12:139–145. Cited on page(s) 45, 46

Bureau of Labor Statistics – BLS. BLS Handbook of Methods, Chapter 10 – Productivity Measures: Business Sector and Major Subsectors, available at `http://www.bls.gov/opub/hom/homch10_e.htm`, accessed 2/14/2012. Cited on page(s) 48

Bruley, M.E. (1998) Is there evidence of a problem? Database search results 1977–1998 (April) ECRI's Health Devices Alerts (HAD) database and FDA's MAUDE database. Presentation made at AAMI/FDA Conference on Medical Device Servicing, Remarketing, and Refurbishing: Is Regulation Needed, Reston, VA, Sept 17–18. Cited on page(s) 23

Centers for Disease Control and Prevention – CDC, (2012) Hand Hygiene in Healthcare Settings - Guidelines, available at `http://www.cdc.gov/handhygiene/Guidelines.html`, accessed on 8/5/2012. Cited on page(s) 12

Cheng, M. and Dyro, J.F. (2004) Good Management Practice for Medical Equipment, in *Handbook of Clinical Engineering*, Dyro J.F. (ed.), Elsevier-CRC Publisher, NY, pp. 108–110. Cited on page(s) 17

Centers for Medicare & Medicaid Services – CMS, (2011) Office of Clinical Standards and Quality/Survey and Certification Group, Clarification of Hospital Equipment Maintenance Requirements (Ref: S&C: 12–07-Hospital), dated 12/2/2011. Available at `http://www.cms.gov/Surveycertificationgeninfo/downloads/SCLetter12_07.pdf`, accessed 6/13/2012. Cited on page(s) 5, 54

Centers for Medicare & Medicaid Services – CMS, (2012a) National Health Expenditure Data, available at `http://cms.hhs.gov/Research-Statistics-Data-and-Systems/Statistics-Trends-and-Reports/NationalHealthExpendData/downloads/tables.pdf`, accessed on 7/22/2012. Cited on page(s) 3

Centers for Medicare & Medicaid Services – CMS, (2012b) State Operations Manual, available at `http://www.cms.gov/Regulations-and-Guidance/Guidance/Manuals/Internet-Only-Manuals-IOMs-Items/CMS1201984.html`, accessed on 8/5/2012. Cited on page(s) 13

Cohen, T. (1997) Validating medical equipment repair and maintenance metrics: a progress report. *Biomed Instr Techn* 31:23–32. Cited on page(s) 46

Cohen, T. (1998) Validating medical equipment repair and maintenance metrics, Part II: Results of the 1997 survey. *Biomed Instr Techn* 32:136–144. Cited on page(s) 46

Cohen, T., Bakuzonics, C., Friedman, S.B., Roa, R.L. (1995) Benchmarking indicators for medical equipment repair and maintenance. *Biomed Inst Techn* 29:308–321. Cited on page(s) 46, 50

Corio, M.R. and Costantini, L.P. (1989) Frequency and severity of forced outages immediately following planned or maintenance outages, Generating Availability Trends Summary Report, North American Electric Reliability Council, Princeton, NJ. Cited on page(s) 23

Cram, N. (1998) Computerized Maintenance Management Systems: A Review of Available Products, *J Clin Eng* 23:169–179. Cited on page(s) 36

David, Y. and Rohe, D. (1986) Clinical engineering program productivity and measurements. *J. Clin Eng* 11:435–443. Cited on page(s) 45

Douglas, K.R. (2012) Science on the sidelines. *Technation* 7:44–49. Cited on page(s) 54

Dyro, J.F. (2004) Accident Investigation, in *Handbook of Clinical Engineering*, Dyro J.F. (ed.), Elsevier-CRC Publisher, NY, 269–281. Cited on page(s) 10

Evidence-Based Medicine Working Group – EBMWG, (1992) Evidence-Based Medicine: A new approach to teaching the practice of medicine, *J Am Med Assoc* 268(17):2420–2425. Cited on page(s) 23, 53

ECRI Institute – ECRI, (2012) Inspection and Preventive Maintenance Procedures, Biomedical-Benchmark™, available at `www.ecri.org/biomedicalbenchmark`. Cited on page(s) 26, 35

Food and Drug Administration – FDA, (2011) Compliance Program Guidance Manual, Chapter 42 – Blood and Blood Products, Inspection of Source Plasma Establishments, Brokers, Testing Laboratories, and Contractors - 7342.002, Center for Biologics Evaluation and Research. Cited on page(s) 8

Fennigkoh, L. and Smith, B. (1989) Clinical equipment management. *JCAHO PTSM Series* 2:5–14. Cited on page(s) 24

Fennigkoh, L. (2004) Cost-Effectiveness and Productivity, in *Handbook of Clinical Engineering*, Dyro J.F. (ed.), Elsevier Academic Press, Burlington, MA, pp. 199–202. Cited on page(s) 46

Frize, M. (1990a) Results of an international survey of clinical engineering departments. Part 1 – Role, functional involvement and recognition, *Med Biol Eng Comput* 28:153–159.
DOI: 10.1007/BF02441771 Cited on page(s) 45

Frize, M. (1990b) Results of an international survey of clinical engineering departments. Part 2 – Budgets, staffing, resources and financial strategies, *Med Biol Eng Comput* 28:160–165.
DOI: 10.1007/BF02441772 Cited on page(s) 45

Furst, E. (1986) Productivity and Cost-Effectiveness of clinical engineering, *J Clin Eng* 11:105–113. Cited on page(s) 45

Furst, E. (1987) Guest editorial - Special Issue Productivity and Cost Effectiveness, *J Clin Eng* 12:99–100. Cited on page(s) 45, 50

Gaev, J. (2010) Successful measures: Benchmarking clinical engineering performance. *Health Facilities Mgmt* 23:27–30. Cited on page(s) 50

Geddes, L.A. (2002) *Medical Device Accidents and Illustrative Cases*, 2nd ed., Lawyers and Judges Pub., Tucson, AZ. Cited on page(s) 10

Healthcare Information and Management Systems Society – HIMSS, (2012) Manufacturer Disclosure Statement for Medical Device Security, available at `http://www.himss.org/ASP/topics_FocusDynamic.asp?faid=99`, accessed on 7/22/2012. Cited on page(s) 7

Hutchins, B. (2006) Suspicion kills national benchmark, *Biomed Instr Techn* 40:327. Cited on page(s) 50

International Standards Organization – ISO, (1999) ISO/IEC Guide 51:1999, 2nd ed., Safety aspects– Guidelines for their inclusion in standards. Cited on page(s) 24, 30

International Standards Organization – ISO, (2000) ISO/AAMI/ANSI standard 14971 - Medical devices – Application of risk management to medical devices. Cited on page(s) 24, 30

International Standards Organization – ISO, (2008) ISO/ANSI/ASQ, Standard 9001:2008 - Quality management systems – Requirements. Cited on page(s) 33, 39

Johnston, G.I. (1987) Are productivity and cost-effectiveness comparisons between in-house clinical engineering departments possible or useful? *J Clin Eng* 12:147–152. Cited on page(s) 45

Kaplan, R.S. and Norton, D.P. (1996) *The Balanced Scorecard*, Harvard Business School Press, Cambridge, MA. Cited on page(s) 41

Kawohl, W., Temple-Bird, C., Lenel, A. and Kaur, M. (2005) *Guide 6 - How to Manage the Finances of Your Healthcare Technology Management Teams*, Ziken International-TALC, St. Albans, Hertfordshire, UK. Cited on page(s) 46

Maddock, K.E. (2006) (Benchmarking) Glass is half full, *Biomed Instr Techn* 40:328. Cited on page(s) 50

Moubray, J. (1997) *Reliability-Centred Maintenance*, 2nd ed., Industrial Press, Inc., New York, NY. Cited on page(s) 23

Nader, R. (1971) Ralph Nader's most shocking expose, *Ladies Home J* 98:176–179. Cited on page(s) 23

National Conference of State Legislatures – NCSL, (2012) Certificate of Need: State Health Laws and Programs, available at `http://www.ncsl.org/issues-research/health/con-certificate-of-need-state-laws.aspx`, accessed on 8/5/2012. Cited on page(s) 14

National Fire Protection Association – NFPA, (2012) *NFPA 99 – Standards for Health Care Facilities*, Quincy, MA. Cited on page(s) 14

Office of Civil Rights – OCR, (2012) Health Information Privacy FAQ, "Is a physician required to have business associate contracts with technicians such as plumbers, electricians or photocopy machine repairmen who provide repair services in a physician's office?" available at http://www.hhs.gov/ocr/privacy/hipaa/faq/business_associates/244.html accessed on 7/22/2012. Cited on page(s) 7

Reason, J. (2000) Human error: models and management. *Br Med J* 320:768–70. DOI: 10.1136/bmj.320.7237.768 Cited on page(s) 31

Ridgway, M. (2004) The great debate on electrical safety – In retrospect, in *Handbook of Clinical Engineering*, Dyro, J.F. (ed.), Elsevier-CRC Publisher, NY, pp. 281–283. Cited on page(s) 5

Ridgway, M.G., Johnston G.I. and McClain J.P. (2004) History of engineering and technology in health care, in *Handbook of Clinical Engineering*, Dyro, J.F. (ed.), Elsevier-CRC Publisher, NY, pp. 7–10. Cited on page(s) 23

Sinek, S. (2009) *Start with Why: How Great Leaders Inspire Everyone to Take Action*. Portfolio/Penguin, New York, NY. Cited on page(s) 51

Smith, A.M. and Hinchcliffe, G.R. (2004) *RCM–Gateway to World Class Maintenance*, 2nd ed., Elsevier Butterworth-Heinemann, Burlington, MA. Cited on page(s) 23

Stiefel, R. (2009) *Medical Equipment Management Manual*, Association for the Advancement of Medical Instrumentation - AAMI, Arlington, VA, pp. 3–5. Cited on page(s) 5

The Joint Commission – TJC, (2012a) Sentinel Event, available at http://www.jointcommission.org/sentinel_event.aspx, accessed on 8/5/2012. Cited on page(s) 1, 22, 53, 54

The Joint Commission – TJC, (2012b) *Comprehensive Accreditation Manual for Hospitals*, Joint Commission Resources, Oakbrook Terrace, IL. Cited on page(s) 5, 26

Wang, B. and Levenson, A. (2000) Equipment inclusion criteria: A new interpretation of JCAHO's medical equipment management standard, *J Clin Eng* 25(1):26–35. Cited on page(s) 24

Wang, B. (2004) Financial Management of Clinical Engineering Services, in *Handbook of Clinical Engineering*, Dyro, J.F. (ed.), Elsevier-CRC Publisher, NY, pp. 188–199. Cited on page(s) 40

Wang, B. (2007) Evidence-Based Maintenance? *24x7 magazine* 12(4):56. Cited on page(s) 23, 53

Wang, B. (2008) Evidence-based Medical Equipment Maintenance Management, in *A Practicum for Biomedical Engineering and Technology Management Issues*, L.R. Atles (ed.), Kendall/Hunt Publishing, Dubuque IO, pp. 219–254. Cited on page(s) 23, 26, 29, 53

Wang, B. (2009) *Strategic Health Technology Incorporation*, Morgan and Claypool Publ., Princeton, NJ. Cited on page(s) 17, 19

Wang, B., Rui T., Balar, S., Alba, T., Hertzler, L.W. and Poplin, B. (2012) Clinical Engineering Productivity and Staffing Revisited: How should it be measured and how can it be used? *J Clin Eng*. DOI: 10.1097/JCE.0b013e31826cc689 Cited on page(s) 1, 26, 44, 46, 48, 54

Wang, B., Rui, T. and Balar, S. An Estimate of Patient Incidents Caused by Medical Equipment Maintenance Omissions, *Biomed Instr Techn*, in press. Cited on page(s) 1, 23, 54

Wang, B., Rui, T., Koslosky, J., and Fedele, J., A comparison of scheduled maintenance methods: OEM-recommended versus hospital-developed. Manuscript in preparation. Cited on page(s) 53

Wang, B., Furst, E., Cohen, T., Keil, O.R., Ridgway, M. and Stiefel, R. (2006a) Medical Equipment Management Strategies, *Biomed Instrum and Techn* 40:233–237. DOI: 10.2345/i0899-8205-40-3-233.1 Cited on page(s) 24, 26, 27

Wang, B., Eliason, R.W. and Vanderzee, S.C. (2006b) Global Failure Rate: A promising medical equipment management outcome benchmark, *J Clin Eng* 31:145–151. Cited on page(s) 44

Wang, B., Eliason, R.W., Richards, S., Hertzler, L.W. and Koenigshof, S. (2008) Clinical Engineering Benchmarking: An Analysis of American Acute Care Hospitals, *J Clin Eng* 33:24–37. Cited on page(s) 1, 37, 39, 41, 44, 46, 48, 50

Wang, B., Fedele, J., Pridgen, B., Rui, T., Barnett, L., Granade, C., Helfrich, R., Stephenson, B., Lesueur, D., Huffman, T., Wakefield, J.R., Hertzler, L.W. and Poplin, B. (2010a) Evidence-Based Maintenance: Part I - Measuring maintenance effectiveness with failure codes, *J Clin Eng* 35:132–144. Cited on page(s) 23, 27, 53

Wang, B., Fedele, J., Pridgen, B., Rui, T., Barnett, L., Granade, C., Helfrich, R., Stephenson, B., Lesueur, D., Huffman, T., Wakefield, J.R., Hertzler, L.W. and Poplin, B. (2010b) Evidence-Based Maintenance: Part II - Comparing maintenance strategies using failure codes, *J Clin Eng* 35:223–230. Cited on page(s) 23, 27, 53

Wang, B., Fedele, J., Pridgen, B., Rui, T., Barnett, L., Granade, C., Helfrich, R., Stephenson, B., Lesueur, D., Huffman, T., Wakefield, J.R., Hertzler, L.W. and Poplin, B. (2011) Evidence-Based Maintenance: Part III - Enhancing patient safety using failure code analysis, *J Clin Eng* 36:72–84. Cited on page(s) 23, 27, 44, 53, 54

Weinstein, R.A. (1998) Nosocomial infection update, *Emerg Infect Dis* 4:416–420. DOI: 10.3201/eid0403.980320 Cited on page(s) 12

Zak, J. The Changing World of Maintenance, (2005) *Proceedings of the 2005 AAMI Annual Meeting and Conference*, Boston MA, Association for the Advancement of Medical Instrumentation, Arlington, VA. Cited on page(s) 23

Author's Biography

BINSENG WANG

Binseng Wang, Sc.D., C.C.E., fAIMBE, fACCE earned Bachelor of Science degrees in both Physics and Electronics Engineering from the University of São Paulo, a Master of Engineering degree from the State University of Campinas, and a Doctor of Science degree from Massachusetts Institute of Technology. He is a Certified Quality Systems (ISO 9001) Auditor and a Certified Clinical Engineer. He started his career in Brazil as a faculty member at the State University of Campinas, where he created the Center for Biomedical Engineering. He also served as the Special Advisor on Equipment to the Secretary of Health of São Paulo state. In the U.S., he worked at the National Institutes of Health and served as Vice President of Quality Assurance and Regulatory Affairs with MEDIQ/PRN Life Support Services, Inc. Currently he is Vice President of Quality and Regulatory Compliance for ARAMARK Healthcare Technologies, where he oversees the nationwide Medical Equipment Management Program that establishes the operating policies and procedures for equipment planning, acquisition, maintenance, retirement, replacement, supplier management, regulatory compliance, risk management, and quality monitoring and improvement. He also ensures the compliance of ARAMARK teams located at over 700 client sites with medical-equipment laws, regulations, and standards. He has traveled around the world providing consulting services to several national governments under the auspices of the Pan-American Health Organization, World Health Organization, Inter-American Development Bank, World Bank, and numerous non-government organizations. Dr. Wang is a fellow of the American College of Clinical Engineering (ACCE) and American Institute of Medical and Biological Engineering (AIMBE), a member of the Health Technology Technical Advisory Group of the World Health Organization (WHO). He received the 2010 AAMI Clinical/Biomedical Engineering Achievement Award.

Printed in the United States
by Baker & Taylor Publisher Services